Energy & Climate

Energy & Climate

How to Achieve a Successful Energy Transition

Alexandre Rojey

Former Director for Sustainable Development at IFP,
Rueil-Malmaison, France

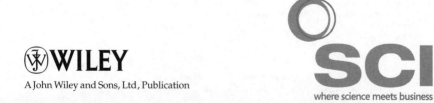

WILEY

A John Wiley and Sons, Ltd, Publication

This work is a co-publication between the Society of Chemical Industry and John Wiley & Sons, Ltd.

This edition first published 2009 by John Wiley & Sons in association with SCI

© 2009 SCI (Society of Chemical Industry)
　　　　14/15 Belgrave Square
　　　　London
　　　　SW1X 8PS
　　　　United Kingdom
　　　　www.soci.org

Authorised translation from French language edition published by Editions Technip, 2008. Translated by the author and Trevor Jones (Lionbridge).

Registered office
John Wiley & Sons Ltd, The Atrium, Southern Gate, Chichester, West Sussex, PO19 8SQ, United Kingdom

For details of our global editorial offices, for customer services and for information about how to apply for permission to reuse the copyright material in this book please see our website at www.wiley.com.

Library of Congress Cataloging-in-Publication Data

Rojey, Alexandre.
　　Energy & climate : how to achieve a successful energy transition / Alexandre Rojey.
　　　　p. cm.
　　Includes bibliographical references and index.
　　ISBN 978-0-470-74427-7 (pbk.)
　　1. Energy development–Environmental aspects. 2. Energy development–Environmental aspects. 3. Power resources–Environmental aspects. 4. Climatic changes–Environmental aspects. I. Title. II. Title: Energy and climate.
　　TD195.E49R65 2009
　　333.79–dc22

　　　　　　　　　　　　　　　　　　　2009007657

A catalogue record for this book is available from the British Library.

ISBN: 978-0-470-74427-7 (PB)

Set in 10.5/13pt Sabon by Thomson Digital, Noida, India.
Printed and bound in Great Britain by TJ International, Padstow, Cornwall.

Contents

SCI Preface

The demanding challenges of the 21st century will only be met if science and business work hand in hand to secure the commercial application of sustainable innovation for the benefit of society. The crucial areas for such application are energy, water and affordable food production for a rising global population, coupled with the urgent need to mitigate environmental pollution.

SCI, an established member-led organisation linking science and business, has selected John Wiley and Sons as its publishing partner for a new range of books designed to stimulate the knowledge exchange that is critical to success.

Each title will summarise scientific research and commercial development in one of the key areas of application, and will seek to establish new paradigms for progress.

This, the first book in the range, examines why we need a sustainable energy system and looks at the spread of technologies that must be harnessed if we are to move successfully away from our current dependence on fossil fuels. Dr Rojey's experience as Director for Sustainable Development at IFP (formerly L'Institut Français du Pétrole) informs a concise and knowledgeable approach to one of the world's most pressing challenges.

Preface

From the discovery of fire and the use of wood for heating and cooking, the implementation of wind energy and fossil energies (coal, oil and gas) up to nuclear energy, one type of energy has always been replaced or succeeded by another. The transition generally occurred before the previous energy supply was completely depleted. There has never been a shortage of energy and, although we experienced serious warnings with the oil crises at the end of the last century, our societies have never really had to worry about its availability. This is no longer true today.

Energy is at the heart of our modern societies. Its importance can now be measured using the yardstick of the debates and controversies it generates. There is fear on the one hand of a future shortage and, on the other hand, concern regarding the damaging effect of its CO_2 emissions in terms of global warming. Concerning the first point, the actual situation is not as black as some people would try to paint it, since hydrocarbon reserves still represent several decades of consumption. As regards the second point, however, we have to admit that urgent action is required. We are at the dawn of an unprecedented scenario: fossil energies are not inexhaustible and cannot be consumed any longer without a significant reduction in their CO_2 emissions. Alternatives are available or emerging, for example the renewable energies, but the conditions of meeting ever-increasing demand at world level, without damaging the environment, are not currently met.

It is difficult to imagine that, from one day to the next, we can suddenly do without hydrocarbons, especially in the fields of transport and chemistry, since there is no substitute which is available in sufficient quantities or which can be used on a large scale under satisfactory economic and environmental conditions. It is even more difficult to imagine that we can maintain the current level of consumption indefinitely without taking action to fight global warming. This last point is vital and represents a global stake for which there is no alternative but to succeed. Relying solely on the progress resulting from technological innovation will not be enough to solve the problem completely. At the same time, our economic and social development model must be rapidly revised, and our consumption modes, and therefore our behaviour, modified. To preserve our planet and our way of life, we must now take a different approach to energy.

Between today and what we think the potential consequences will be tomorrow if nothing is done, we must conceive without delay the most balanced and logical path which will allow future generations to evolve and continue their own development under acceptable economic, social and environmental conditions. The aim is not to set the two objectives against each other – energy availability and use versus the fight against global warming – but, on the contrary, to reconcile them. This is where the immediate implementation of a 'controlled' energy transition takes its full meaning and importance. In order to make this transition a success, all players must be convinced that there is a real need for urgency and that consequently, there is an unprecedented need for all parties to act together and implement the means required. We must also accept that there is no single 'miracle' solution in the immediate future and that several complementary solutions must be set up, without any preconceived ideas and without criticising one compared with another. The role of the public authorities is to motivate and control this movement through financing, regulation and other incentives. The economic players, especially companies, must integrate it into their development strategy. For companies, the expression 'citizen company' has never had as much meaning and importance as it does today. Emphasis must be placed on research, whether public or private, to develop the technical solutions we will need and this requires the allocation of finance and human resources. It is the duty of all individuals in their daily lives to reconsider their actions and radically change their consumer behaviour. Energy savings must now be our main concern. Lastly, and not the least of the success factors, all players must take concerted action and not act individually.

As described in this book, the economic and social components of all sectors of activity are concerned with the stakes relating to availability of energy and environmental issues. The solutions and developments known or to be expanded may sometimes be common and sometimes specific. Some sectors of activity have a technological lead over others. This is the case with housing, for example, which has a number of immediately operational solutions and a relatively clear roadmap. The difficulty is not so much in the existence of technologies and materials but in the conditions of their deployment, in terms of the level of investment required and, also, the slow rate of renewal of dwellings – a house or building is built for a period of 50 to 100 years. Nonetheless, there is still room for progress even in the housing sector, and research must be conducted accordingly.

All countries in the world are equally concerned. We are faced with a global problem. Some technologies can be deployed everywhere, others will be specific to countries or geographical areas and will depend on choices related to local weather conditions, the economic and social situation or quite simply the availability and nature of energy sources.

Making the energy transition a success is undeniably everyone's concern. Every person has a role to play at his or her own level. IFP, a public research and training organisation whose mission is to develop the transport energies of the twenty-first century, is firmly committed to making its contribution.

On the strength of its achievements, with its men and women, their experience, their skills, their know-how and, above all, their outstanding capacity for innovation, IFP intends to actively and efficiently take part in a controlled transition in the fields of energy, transport and the environment.

This book, which extends far beyond IFP's scope of action, is as complete as possible and provides a vision and global expertise of the energy transition issue. Obviously some of the subjects discussed are included in the various fields of research covered by IFP. The five strategic priorities of IFP can be summarised as follows: capture and store CO_2 to combat global warming; diversify fuel sources; develop clean, fuel-efficient vehicles; convert as much raw material as possible into energy for transport; and push back the boundaries in oil and gas exploration and production. While the analysis and proposals in this book reflect the author's point of view, the description of the stakes and solutions required in each of these priorities are globally in line with the IFP vision. Making the energy transition a success is an objective, as much as an obligation, shared by all IFP scientists in their daily work. Apart from the analysis proposed, this book is above all a warning and a strong incentive to act without delay. IFP fully agrees with this need for action.

Olivier Appert
IFP Chairman and CEO

About the Author

Dr Alexandre Rojey was, until recently, Director for Sustainable Development at IFP (Institut Français du Pétrole) in France, responsible for new developments in the energy sector such as the use of hydrogen, long term energy supply and issues related to global climate change including CO_2 capture and storage. He is also the Chairman of CEDIGAZ, an international association in the area of natural gas, and an active member of IDéEs (for Innovation, Développement durable, Energie et Société), a think tank concerned with energy and sustainable development in France. He graduated from the Ecole Centrale de Lyon, one of the most prestigious French engineering colleges and has contributed to more than 70 publications including six books. He also holds over 100 worldwide patents.

Introduction

Energy plays a central role in our society. All our sectors of activity, housing, transport, industry and agriculture depend on it.

The significant technological mutations in the field of energy have led to major changes in the economic and social operation of the world in which we live. This is illustrated, for example, with the industrial revolution, marked by the advent of the steam engine and coal as energy source, development of automotive and air transport, related to large-scale use of oil in the twentieth century and the increasing role played by electricity in our economic system.

Currently, however, increased energy consumption is no longer solely associated with the notion of progress, but also with a certain number of threats which weigh on our society and our environment.

The risks of depletion of fossil energies, above all oil, represent an initial factor for concern. The imminence of peak oil production is mentioned repeatedly. Tensions over oil supplies, induced by increasing demand, cause continued price instability and combine with political factors to create the threat of crisis.

The impact of energy production from fossil fuels on the environment is also becoming a matter of growing concern. In addition to the risks for the environment on a local scale, we are now faced with the danger of global warming caused by CO_2 emissions.

The solution may seem to be straightforward: simply stop using fossil energies and replace them by energies which do not display the same disadvantages, namely nuclear energy and the renewable energies, thereby moving into the 'post-oil era'.

Unfortunately in practice, on a global scale, these alternative energies cannot be substituted rapidly and massively for the fossil energies, which will continue to represent a substantial share of the energy consumed for many years to come. Also, they are not risk-free, either in terms of safety as regards nuclear energy or in economic terms as regards the renewable energies. Moving from the current situation to a sustainable energy system cannot be done by waving a magic wand. This move will involve, in particular, radically changing our habits as well as our energy production and consumption structures.

We must therefore plan a **transition** to avoid, first, a major crisis in energy supplies and, secondly, a climate change with catastrophic consequences. The paths to be taken to ensure the success of this energy transition will be discussed in the remainder of this book:

- Chapter 1 describes how energy is currently consumed in the context of globalisation, which favours a sharp rise in demand.
- The threats for consideration in this consumption model are analysed in Chapters 2 and 3. The analysis demonstrates that the current model is not sustainable.
- Chapter 4 deals with the conditions under which the energy transition must be undertaken. This analysis results in an action plan based on four points.
- The actions to be carried out according to these four points are described in greater detail in Chapters 5–8. They are a collection of innovations which will be described and summarised in a table. Technical solutions alone, however, are not sufficient. A change in patterns of behaviour and the statutory context is also an integral part of the strategy proposed.
- Chapter 9 describes the forthcoming perspectives, based on an analysis of the future contribution of these various solutions in solving the problems raised.

An acceptable development scenario is within our reach, but at the price of an effort which must not be underestimated. A substantial change is necessary. It concerns not only the energy sector, but also our entire economy and our society. The commitment of all citizens, companies and governments will be crucial to successfully implement the actions required. In view of the imminence and gravity of the dangers facing us, this change is urgent: we must act with determination and without delay.

1

Energy in a Globalised World

The fundamental role of energy in the economy

Energy is omnipresent in the economy and plays a fundamental role in all fields of activity, whether it is in industry, the residential and tertiary sectors or transport[1].

The development of coal accompanied the discovery of the steam engine, leading to the first industrial revolution in the eighteenth century. The use of coal, followed by other fossil energies (oil, natural gas) to drive machines, allowed the incredible development of industry, up to the present time.

The industrial revolution was also marked by the spectacular development of transport. The use of the steam engine to drive trains and ships resulted, from the start of the industrial era, in the creation of rail and sea networks across the globe. At the start of the twentieth century, plentiful supplies of oil, easy to store in liquid form and relatively cheap, sparked the rapid growth of road and air transport.

Energy also caters for heating and air-conditioning requirements. It is essential to operate all household, office (computers) and communication (media, telephone, internet) equipment. Industry is totally reliant on it.

Without this continuous supply of energy, society would grind to a standstill and, in this respect, our modern economies are particularly vulnerable.

The world economy demands increasing quantities of fossil fuels to meet its energy requirements. The development of road and air transport, directly related to oil consumption, is the most flagrant illustration. The

---

[1] Some notes concerning the units used and the physics involved are provided in Appendix 1.

Energy and Climate: How to achieve a successful energy transition Alexandre Rojey
Copyright © 2009 Society of Chemical Industry

distribution of relatively inexpensive means of transport on a worldwide scale has encouraged globalisation of the economy which, in return, increases energy consumption.

The alternative energies, which have developed more recently, are still far from being on an equal footing. Nuclear energy is largely considered throughout the world as presenting a number of risks: risks of diversion for military purposes, accidents, and waste storage. The renewable energies, supposed to eliminate these various constraints, are developing only slowly and their actual ability to replace the current energies is sometimes questioned.

Current predominance of fossil energies

According to the 2008 issue of the *World Energy Outlook* published by IEA, in 2006, the world supply of primary energy (from oil, natural gas, coal, nuclear sources and renewable energies) amounted to 11.7 billion tonne oil equivalent [1].

One tonne oil equivalent (toe) represents the energy obtained through combustion of 1 t of oil. Even though the increasing electrification of our economy also brings other units into use, such as the kWh and the MWh (1 MWh = 0.086 toe), the widespread use of the toe as an energy unit shows that oil remains the reference energy.

Primary energy is that available before any conversion, except for energies which cannot be exploited directly (Figure 1.1). Hydroelectricity and nuclear energy are therefore considered as primary energies. Electricity produced from nuclear power is assigned an equivalence

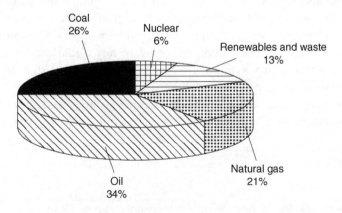

Figure 1.1   Sources of primary energy (2006 figures – Source: IEA)

coefficient based on 0.33 efficiency (0.261 toe for 1 MWh of nuclear electricity).

Primary energy is then converted into secondary energy, into a form which can be used by the consumer (production of fuels in the refineries and electricity in the power stations).

The final energy is consumed by the user after the transport and distribution process.

Currently, fossil energies are by far the most widely used, providing a little over 80% of the world supply of primary energy [2]. Oil still represents the most important share (34%), followed by coal (26%), natural gas (21%), renewable energies (13%) and lastly nuclear power (6%).

Most of the renewable energy supply comes from biomass (nearly 80%). Hydroelectricity represents only 16% of this supply and the other forms of renewable energy (including wind and solar) just 4%, i.e. about 0.5% of the total primary energy consumed.

After a fast development phase in the 1970s, the proportion of nuclear power stagnated and even dropped, following the concerns raised by the accidents at Three Mile Island and especially Chernobyl, which exacerbated public distrust. This trend was amplified by the oil counter crisis, which reduced the financial attraction of nuclear power for a long period of time.

At the same time, extensive development in road and air transport generated a high demand for oil, virtually the only source of the fuels used for which there is no immediate substitute at the required scale.

In this context, the share of fossil energies used has remained more or less stable. The share of oil has dropped slightly, with a corresponding increase first in natural gas and more recently coal, which has made a remarkable come-back over the last five years. Although 'post-oil' is often mentioned, making substantial changes to the distribution of the various primary energy sources will involve considerable time and effort.

Uses of energy

Consumption of final energy worldwide is distributed as follows between the main sectors of activity (Figure 1.2):

- The residential (dwellings), tertiary (offices and service activities) and agricultural sectors represent 36% of the global energy demand.

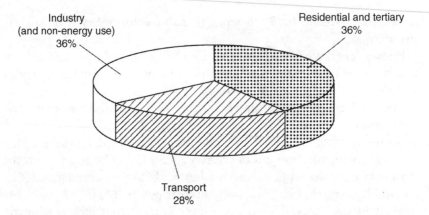

Figure 1.2 Distribution of the global energy demand (final energy consumption) (Source: IEA)

- Transport (road, air, rail, sea) accounts for a share of 28%.
- Industry consumes 36% of the final energy

In the residential and tertiary sectors, energy is mainly consumed for heating and the supply of hot sanitary water, as well as for air-conditioning. Much of this demand is therefore seasonal. Fossil energies, which are easily stored, are ideal for this type of irregular demand. This is less true of the other forms of energy, apart from hydraulic. In particular, energies exploited directly as electricity are difficult to store. Solar contributions are intermittent and offset with respect to the demand, in particular as regards heating requirements.

Road and air transport poses a highly critical problem, considering the rocketing increase of the demand in these sectors and the limited possibilities of turning to substitute fuels.

Improved standards of living are accompanied throughout the world by a rapid rise in the number of vehicles and increasing use of air transport.

Simultaneously, goods transport by road has developed considerably at the expense of rail traffic. This trend is related to the undeniable advantages of road transport in terms of shipment flexibility and implementation of 'just in time' production lines, with minimisation of stocks. It has also been favoured by the state, which has paid for most of the road infrastructure costs and done little to encourage rail freight transport.

Between 2000 and 2050, road passenger traffic is expected to increase by 150% and goods traffic by 200%. Air traffic, for passengers and goods, is also increasing sharply, at an annual rate of about 5%. This trend will inevitably have an impact on oil demand.

In industry, energy is consumed mainly as heat, often at high temperatures, to convert raw materials (metallurgy, iron and steel, glass, cement, etc.) and as mechanical energy to drive machinery.

Electricity holds a special place amongst the various types of energy. It represents 16% of the final energy demand and about 37% of the primary energy consumption, the difference between these two figures being due to electricity production efficiency, which is taken into account when calculating the primary energy consumption.

Electricity production is rising steadily at an average rate of 3% per annum and its share in the total energy consumption is rising regularly throughout the world [3]. By 2030, according to the IEA reference scenario, this share is expected to change from 37% to 41% in the primary energy balance and from 16% to 21% in the final energy balance.

Electricity consumption is rising in particular in the residential and tertiary sectors, due to the growing number of devices (household appliances, media, etc.) and more intensive use of electricity for air-conditioning and heating.

Special vigilance is required in the transport and electricity production sectors, whose fossil energy consumption is rising at an alarming rate.

Restricted number of suppliers

The distribution of fossil fuel reserves throughout the world, expressed in billions of toe, is represented on the map in Figure 1.3.

World reserves of hydrocarbons, oil and natural gas, which provide more than half of the primary energy supply, are unequally distributed, mostly located in regions far from the main consumption areas.

Consequently, more and more of these hydrocarbons must be imported by the consumer countries. The fact that some of the oil and gas reserves are located in unstable regions of the world creates geopolitical risks.

In addition, the problem of security of supplies is aggravated by the limited number of players controlling a major proportion of the reserves (Saudi Arabia for oil, and Russia, Qatar and Iran for natural gas).

The ambition of the producing countries to control the market, which resulted in particular in the creation of OPEC[2], has only been partially achieved. Nonetheless, the current situation is not especially favourable for the consumer countries.

---

[2] Organisation of Petroleum Exporting Countries.

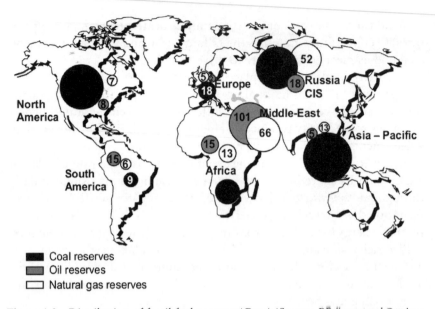

Figure 1.3   Distribution of fossil fuel reserves (Gtoe) (*Source*: BP Statistical Review 2008)

In the sector of hydrocarbons, the producing states are currently generally reluctant to open their oil and gas fields to the big international companies. This is the case for most of the OPEC members: Saudi Arabia and Kuwait are completely closed. Outside of OPEC, this is also true of a country like Mexico. Recent times have been marked, in several producing countries, for example Russia, Venezuela, Iran and Qatar, by a greater desire for national control, resulting in particular in renegotiation of contracts signed with the international companies.

The political situation of some of the countries in the Middle East, which contains 57% of world oil reserves and 40% of world natural gas reserves, is particularly unstable. Oil production in Iraq has dropped to about 2 million barrels a day, due to the current conflict, a figure much below its actual production capacity. In Iran, the conflict with the international community over the nuclear issue is delaying investments and new developments in the hydrocarbon sector. Saudi Arabia itself seems to be in a delicate situation, having to cope with a declining economy and the threat of terrorism [4]. Oil is often a source of conflict and oil revenue can often be diverted to finance wars.

Reserves of natural gas are less concentrated; Russia holds a leading position with one-third of the global reserves, and is one of the main exporting countries alongside Canada, Norway, Algeria, the Netherlands and Indonesia.

However, long distance transport of natural gas through gas pipelines or as liquefied natural gas (LNG)[3] relies on large-scale infrastructures and therefore major investments. The mutual dependency between supplier and consumer, induced by the existence of gas transport infrastructures, represents a safety factor. The recent tensions between Russia and the European Union, due in particular to its conflict with Ukraine, have nevertheless demonstrated the risks inherent in excessive dependency on a dominant supplier.

The distribution of coal reserves is more favourable to the large consumer countries. Due to their abundance, coal has not suffered the tensions and price rises inflicted on oil and natural gas. This explains the strong come-back of coal observed over recent years.

This come-back nevertheless involves major environmental risks, especially regarding CO_2 emissions[4], as discussed in Chapter 7.

Energy and globalisation of the economy

Access to ever-increasing quantities of oil has favoured the multiplication of exchanges between the various parts of the world, whether for people with the sharp increase in air travel, or goods with the development of road and air freight.

This development of transport and exchanges, combined with the tremendous progress made in the field of telecommunications, has been a major factor in globalisation of the economy.

The relatively low cost of energy has closed the gap between the continents and transformed the world into the 'global village' we now know. In this context, European consumers eat fruit and vegetables in winter grown in the southern hemisphere (from Australia, New Zealand and Chile).

The search for lowest-cost production sites has accelerated the development of goods transport and led to a sharp rise in energy consumption by emerging countries.

Simultaneously, liberalisation of the economy and deregulation of the energy sector in most industrialised countries have created the conditions

---

[3] LNG is transported by tankers in liquid phase at low temperature.

[4] Carbon dioxide (CO_2) is emitted during combustion. The fossil fuels, coal, oil and natural gas, represent the main human sources of CO_2 emissions.

for increasingly open competition. This situation has fostered an outbreak of concentrations (mergers and acquisitions) designed to make the newly created industrial groups large enough for the scale of the targeted markets.

This progressive concentration of industrial companies has given greater power to the international companies operating in the energy sector [5] and reduced the role of the states.

These concentrations first took place in the oil sector with the Exxon-Mobil, BP-Amoco-Arco, Chevron-Texaco, Conoco-Phillips and Total-Fina-Elf mergers.

They also affected the electricity production sector, with numerous mergers in Europe strengthening the position of some major players such as E.ON, RWE, Suez and EDF.

The situation has not yet stabilised in this sector and changes are still taking place, as illustrated by the recent merger between Suez and Gaz de France and the acquisition of Endesa by ENEL, following E.ON's takeover bid.

Opening the gas and electricity networks to third parties represents another means of increasing competition. The networks are managed independently of the energy suppliers which use them at the same time. In Europe, the internal energy market must now be open and competitive. This involves separating the transport networks from the production means, under the control of the national regulation authorities.

Globalisation of the economy has led to greater dependency on oil, due to the ever-increasing demand for petroleum fuels and more generally fossil fuels. We have recently seen that the global economy could cope reasonably well with an increase in the cost of energy. In contrast however, an interruption, even temporary, in oil supplies would cause a major crisis.

In this globalised economy, investments and the decision processes depend more and more on the international groups in the energy sector, at least during the energy conversion and supply stages. In addition, since the energy policies within the European Union are becoming increasingly integrated, fewer decisions can be made at a purely national level.

The increasing role of the emerging countries

Following the globalisation of the economy, an increasing proportion of the industrial activities, in particular those consuming large quantities of energy, has been transferred to the emerging countries, China, India and Brazil.

Delocalisation of much of the production to these countries, together with the fast development of their economies, has resulted in a sharp upsurge in their energy consumption. This trend, particularly striking in China, tends to exacerbate tensions.

There are numerous consequences. Not only does development of the emerging countries intensify the rate of growth of the world energy demand, it also has implications for the distributions between the various energies consumed. The growth of China, for example, has quickly driven up the demand for coal.

This situation also has geopolitical consequences. Now having become a major economic power and having to import an increasing share of its oil, China is seeking to secure its supplies by strengthening its political influence over the producing countries.

The influence of the Western economies, in particular Europe, is therefore reduced in the future trends of the energy sector.

Future outlook

Globalisation is frequently accused of involving unfair exchanges and creating unemployment. It cannot be denied, however, that it has favoured rapid development in countries such as China or India. It is in this context that we have been witnessing exceptional economic growth on a worldwide scale.

In the energy sector, this growth is faced with two major problems: the dependency on energy supplies, and the impact of energy consumption on the environment.

Escalation of road and air traffic generates high dependency on oil supplies. The increasing level of urbanisation reinforces this dependency. This situation makes the world economy highly vulnerable to an interruption in supplies, since no immediate substitution solutions are available. In addition, the search for the lowest costs and the highest competitiveness on a global scale does not favour long-term investments.

The recent financial and economic crisis is an illustration of the growing risks we are facing.

It is also becoming more and more difficult for the environment to withstand the impact of energy consumption. The damage caused is not included in the production costs.

Consequently, despite a certain degree of progress recorded in the richest countries, pollution affects the entire world on an increasingly global scale.

Climate change due to CO_2 emissions caused by the use of fossil fuels has become one of the main risks for the future.

This change, which affects the entire planet, is also a symptom of globalisation.

Unfortunately, globalisation, which has been until now mainly economic and financial, has been of little help in collective resolution of the problems affecting our planet. Energy and the environment are global issues and the solutions to be implemented must be managed through international governance.

The threats faced will be analysed in the next two chapters. They show that, despite its successes, the current economic model is not sustainable and that radical changes in lifestyles as well as in energy production and utilisation conditions will be necessary.

2

Growing Risks Ahead

The risks associated with energy consumption

The continuous rise in energy demand generates increasing risks not only for our economy, but also for our entire planet as a living, habitable system.

Energy production and consumption place increasing pressure on the environment. In addition to the problems of environmental protection on a local and regional scale, we are now faced with the threat of climate change.

At the same time, energy demand is now so high that it is becoming increasingly difficult to guarantee the corresponding supply. We are therefore faced firstly with the risk of not being able to meet the future energy demand, and secondly with the threat of major and irreversible damage to our environment. These threats have arisen in a context of global economic growth which does not encourage people to adopt an alarmist attitude. In view of the progress made, the possibility of a change is difficult to accept. However, it will only be possible to remove the impending dangers through a clear awareness of the risks involved.

The growing demand

Energy demand is growing steadily to meet the requirements of an expanding global population with an improving standard of living. Figure 2.1 illustrates the problem [2].

In 2006, the total consumption of primary energy was 2.4% higher than 2005. In its reference scenario, the IEA forecasts an average annual

Energy and Climate: How to achieve a successful energy transition   Alexandre Rojey
Copyright © 2009 Society of Chemical Industry

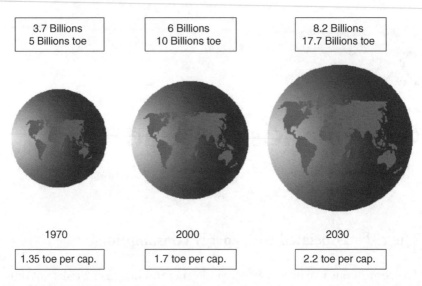

| 3.7 Billions | 6 Billions | 8.2 Billions |
| 5 Billions toe | 10 Billions toe | 17.7 Billions toe |

| 1970 | 2000 | 2030 |
| 1.35 toe per cap. | 1.7 toe per cap. | 2.2 toe per cap. |

Figure 2.1 Growth in population and energy demand (*Source*: IEA). Reproduced with permission from *Energie & Climat: Réussir la transition énergétique* by Alexandre Rojey, Éditions Technip, Paris, 2008

growth of 1.6% by 2030, which represents a 50% increase in demand compared with today.

More than two-thirds of this growth will come from the emerging and developing countries.

During the twentieth century, worldwide energy consumption was multiplied by a factor of about ten, representing an average global growth of 2.3% per annum.

The current growth rate lies between 2% and 2.5%. This growth in demand is primarily due to demographic changes. The world population, just above 2 billion 50 years ago and about 4 billion 30 years ago, has now reached a figure of 6.5 billion people. It is expected to reach 8.2 billion in 2030, and then 9.1 billion in 2050, despite the predicted slowdown in birth rate.

The rising standard of living also results in higher energy consumption. The average energy consumption of each inhabitant is expected to grow from 1.7 toe in 2000 to 2.2 toe in 2030, as shown in Figure 2.1, based on the IEA reference scenario.

The combination of these two factors, higher population and higher standard of living, steadily drives up the energy demand. The primary energy demand is therefore expected to increase from 10 billion toe in 2000 to 17.7 billion toe in 2030 in the reference scenario. If the growth in population and energy consumption per inhabitant should then slow down, the primary energy demand could reach a figure of 23 billion toe in 2050. In other words, for a population of 9.1 billion in 2050, this would mean a figure of 2.5 toe per inhabitant. With higher growth assumptions, this demand could be far greater. We should on the contrary be aiming for a reduction in the primary energy demand per inhabitant; an alternative scenario based on an energy consumption of 1.3 toe per inhabitant in 2050, will be discussed below (see Chapter 9).

Unequal access to energy

While the power produced by a man working is in the region of 100 W, the average power currently available continuously to every inhabitant on the planet is 2.3 kW, which corresponds to an average energy consumption of 1.7 toe per annum.

There are large disparities, however. Every citizen in the USA consumes 8 toe per annum, i.e. twice and ten times as much as European and Chinese citizens, respectively.

With 5% of the world population, the USA uses almost 25% of the primary energy produced in the world and 50% of the automotive gasoline consumed by cars.

In the field of automotive transport, despite rapid progress, the Chinese only have 20 cars per 1000 inhabitants, whereas the Europeans have 600 cars and the Americans 800 cars.

Unable to meet their energy requirements, poor populations are forced to use wood resources, which are often very limited, for cooking and heating purposes, thereby accelerating deforestation. When exploitation of biomass leads to desertification, it is obvious that we can no longer speak of 'renewable' energy.

These inequalities have existed for so many years that in the most privileged countries they are regarded as quite normal. Nevertheless, the rapid development of large emerging countries such as China and India demonstrates that this type of situation can change quickly. Other countries, in particular those in Africa, will need energy to allow their economic growth. It is essential to consider these fundamental requirements which, in the current situation, imply higher energy demand.

Risks for the long-term energy supply

The continuous growth in demand raises the question of the availability of energy in the long term, in sufficient quantities to avoid a major crisis,

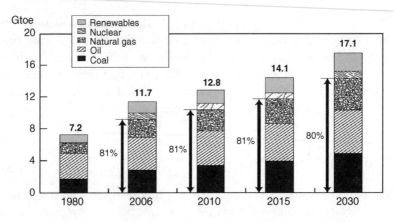

Figure 2.2   Trend in primary energy supply (*Source*: IEA)

bearing in mind that most of the energy comes from resources which, by definition, are limited. As already mentioned in the previous chapter, energy supplies currently rely mainly on fossil fuels, which represent slightly over 80% of the primary energy consumed worldwide. According to the IEA reference scenario, this proportion is expected to remain stable until 2030 (Figure 2.2). It is true that fossil energy resources will not be depleted in the short term. If consumption continues at the current rate, the proven reserves represent about 40 years for oil, 64 years for natural gas and almost 150 years for coal [8, 84, 86].

Uncertainties nevertheless affect the energy supplies, oil in particular. Reserves are being consumed faster than they are renewed: currently, only one-third of the proven reserves consumed on a global scale is renewed every year. Oil production will therefore inevitably reach a ceiling before starting to decline. This point is examined in more detail in Chapter 7.

Greater vulnerability of consumer countries

Considering the distribution of fossil fuel reserves, we see that the industrialised countries are highly dependent on imports of fossil fuels, especially oil. Most hydrocarbon reserves are concentrated in the Middle East and, over the years, the industrialised countries have become increasingly dependent on oil imports.

This is the case for the USA whose oil imports are rising dramatically: 65% of their oil requirements are currently imported.

Concerning natural gas, since the quantities imported from Canada are expected to decline, the USA must import more and more LNG.

The European Union's energy dependency is particularly high as regards oil (85% of oil imported, without including Norway, 65% including Norway). It is close to 50% for natural gas (57% without Norway, 43% including Norway) and 40% for coal.

In Asia, Japan is totally dependent on imports for its consumption of oil and natural gas, supplied as LNG. China's spectacular economic growth is creating increasing tension on the energy markets, as growth in Chinese demand has created a surge in imports. The dependencies of China and India are tending to increase as their economies improve; these two countries deploy the infrastructures and investments designed to favour energy supplies, primarily oil. In addition, China relies heavily on its coal reserves, which has serious implications for the environment. These constraints could encourage this country to further diversify its energy sources, placing more emphasis on natural gas in particular.

In this context, the dependency on energy imports represents an increasing risk factor for the consumer countries and a factor of instability. Whenever deliveries from an exporting country are stopped or simply reduced, the shock waves are felt throughout the world, often inducing sudden price rises.

Dependency of transport on oil

The transport sector is particularly vulnerable. Road and air transport, which relies on liquid fuels derived almost exclusively from oil, accounts for consumption of more than half of the oil produced worldwide.

Road transport is 97% dependent on oil. The rapid increase of passenger and goods traffic will inevitably add to the oil demand, unless substitution solutions are rapidly set up, which is no easy task. Liquid fuels provide a means of storing large quantities of energy, in a form which is easy to handle and, as we will see in Chapter 6, it is difficult to find substitutes offering comparable advantages.

The dependency of air transport is also a matter of concern in view of the higher traffic, which is expected to double every fourteen years. The share of aviation in world consumption of fuels intended for transport, which has already risen from 10.7% around 1990 to 13.5% in 2000, is expected to increase even further in the future. At the current time, air transport is totally dependent on the supply of kerosene produced from oil.

Increasingly dependent on road and air transport, international trade would be completely paralysed if oil supplies were cut.

Effects of energy consumption on the environment at local and regional levels

Energy consumption results in emissions of pollutants. Local pollution, characteristic of urban areas or caused by emissions from an industrial plant (factory, cement works, refinery, etc.), is converted into regional pollution through accumulation of pollutants in the atmosphere.

Pollution becomes global when it concerns the entire planet, a situation we are facing with the greenhouse gas effect.

Local pollution in an urban environment is due mainly to transport. The main pollutants discharged by automotive vehicles are carbon monoxide, unburnt hydrocarbons, nitrogen oxides and particulates.

High contents of nitrogen oxides and unburnt hydrocarbons in the atmosphere favour photochemical reactions which occur in the presence of high solar radiation, and therefore mainly in summer. These reactions result in the formation of ozone and other oxidising compounds (hydrogen peroxide, aldehydes, peroxyacetylnitrate or PAN). The ozone which therefore forms in the lower atmospheric layers (tropospheric ozone) is irritating and noxious to breathe. It has an adverse effect on health, especially on the respiratory system, as well as on the environment since it attacks vegetation, thereby acting as a pollution indicator, unlike the ozone present in the higher atmospheric layers (stratospheric ozone) which protects against excessive ultraviolet radiation and which must be preserved.

The fine particulates emitted in particular by diesel engine vehicles represent a source of pollution which is especially serious in urban areas. Particulates with diameters less than $10 \, \mu m$ (PM_{10}) are taken into account when analysing atmospheric pollution. Particulates with diameters less than $2.5 \, \mu m$ ($PM_{2.5}$) seem to be the most dangerous. The finest particulates, with diameters between $0.01 \, \mu m$ and $0.1 \, \mu m$, penetrate deeply into the respiratory tracts, into the pulmonary alveoli, from where they are only eliminated very slowly provided that exposure to the pollution is not permanent, causing pulmonary and cardiac risks. Pollutants adsorbed on the particulates also penetrate into the organism via this pathway.

In addition, some aromatic organic compounds (benzene), polycyclic compounds (PAHs) and aldehydes (e.g. formaldehyde) emitted by automobiles are potentially carcinogenic. In the industrialised countries (USA, European countries, Japan), significant progress has been made in the reduction of pollutant emissions, through the introduction of increasingly stringent standards for new vehicles.

Pollution of large urban centres nevertheless remains difficult to control due to the constant rise in the number of vehicles, despite the technical advances made. This is especially true in the developing countries.

Significant pollution remains at regional level, due in particular to the presence of sulphur in the fossil fuels (coal, heavy fuel oil). Emissions of sulphur and nitrogen oxides contribute atmospheric pollution, leading to acid rain which is harmful for forests and lakes. The atmosphere is also polluted by soot and tar emitted by vehicles or industrial facilities, carbon monoxide resulting from incomplete combustion, unburnt hydrocarbons and volatile organic compounds (VOCs) from the evaporation or incomplete combustion of fuels.

In Asia, industrial pollution caused by development of the economy produces, between April and October, a vast brown cloud consisting of sulphur-containing aerosols mixed with carbon monoxide, ozone, nitrogen oxides, soot and dust. This cloud reduces solar radiation and rain by 20% to 40%.

Progress is also underway in this area, both as regards the technical solutions and applicable regulations. The Gothenburg protocol, approved by the United Nations in 1999, plans a 75% reduction in sulphur dioxide (SO_2) emissions and a 49% reduction in nitrogen oxide (NO_x) emissions compared with the 1990 levels [6].

Amongst the problems encountered, sea pollution related to the transport of crude oil receives high media coverage. The consequences of oil spills caused by the accidental sinking of tankers are highly visible, but represent only a relatively small fraction of all hydrocarbon spills at sea. The number of oil spills is in fact steadily decreasing.

The number of accidental oil spills at sea has been divided by 10 over the last 30 years, whilst since 1980 maritime oil traffic has increased by 80%. In addition, the accelerated introduction of double hull tankers has reduced even further the risk of oil slicks. In contrast, deliberate discharges due in particular to tank cleaning, which are difficult to observe and punish, continue and represent a major source of oil pollution at sea.

The risks for the environment are not restricted to fossil fuels. The development of nuclear energy also raises a certain number of problems. The risks involved are not easy to estimate, since they are generally related to accidental phenomena rather than to normal operation. Consequently, the level of danger for the environment and the public is a subject of heated debate. It is related to the management of radioactive waste and to accidental leakage of toxic products throughout the production chain.

A constant effort is therefore required to protect the environment. The actions required to fight against climate change must not detract from the measures to be taken to protect the environment at a local level.

Risks for the environment at world level

Globalisation, which favours delocalisation of production activities, also delocalises pollution. Driven by a search for the most competitive production conditions, firms choose to locate in geographic sectors where environmental regulations are most lax.

Production sites concentrating polluting activities, sometimes called 'pollution havens', are set up in the developing countries. The effect of globalisation is therefore to reduce pollution in the richest countries and aggravate it in the developing countries [6].

Pollution tends to spread to a larger and larger scale; from local, it becomes regional, and then global.

Formation of the 'ozone hole' over the pole, due to destruction of tropospheric ozone by the halogenated compounds formerly used in particular as refrigerating fluids, is a case of global pollution. The problem was solved by an international agreement banning the use of these fluids [7].

Over the last ten years, the issue of climate change has become one of the most worrying. Directly related to the considerable emissions of CO_2 resulting from the use of fossil fuels, it now represents the main global pollution.

An international agreement between most of the countries concerned must be set up to cope with this situation. The degree of difficulty in reaching this type of agreement obviously depends on the degree of effort required.

Growing threats

The economy's high dependency on energy consumption also generates two major risks:

• Interruption in the supply of hydrocarbons

The world economy has a certain amount of flexibility with respect to oil supplies, but room for manoeuvre remains limited, especially when no short-term substitution solution is available, which is the case in particular for road and air transport. The impact of a crisis must not be under-estimated. Not only would such a crisis limit transport, it could also bring the entire economy to a standstill. In addition, development of the poorest countries would be irredeemably jeopardised, further increasing poverty and tensions in the world.

- **An environmental disaster**

Despite a certain degree of progress, the continuing atmospheric, sea and soil pollution could have dramatic consequences on health, water quality and agricultural production.

Currently however, the greatest and most immediate risk concerns the climate change caused by human sources of greenhouse gas emissions.

The next chapter analyses the effects of climate change, which jeopar-dise the very future of our Earth.

3

The Threat of Climate Change

A major risk for the planet

Although the effects on the environment of greenhouse gas emissions related to energy consumption have only been felt quite recently, their consequences are potentially disastrous.

When fossil fuels are burned, the carbon extracted from the ground as oil, gas and coal is discharged into the atmosphere as CO_2. As a result, the CO_2 concentration in the atmosphere increases progressively. Despite being present in relatively low amounts, the CO_2 modifies the transparency of the air with respect to infrared radiation, acting like the glazing on a greenhouse.

This leads to global warming, a process which has already started and which can be expected to accelerate if the necessary measures are not taken in time.

As far back as the end of the nineteenth century, the Swedish chemist Arrhenius had predicted the consequences of an increase in atmospheric CO_2 content on the increase in average temperature on the surface of the earth. However, the risks involved have only been truly understood recently.

Global warming now receives extensive media coverage and is discussed in numerous international summits. The actions initiated so far nevertheless fall well short of what is actually required.

Energy and Climate: How to achieve a successful energy transition   Alexandre Rojey
Copyright © 2009 Society of Chemical Industry

The greenhouse effect

The CO_2 emitted in the atmosphere behaves like a greenhouse gas, according to the mechanism shown in Figure 3.1. The atmosphere is transparent to the incident solar radiation transmitted in the visible light spectrum, but some of the solar energy received by the Earth is reflected back as infrared radiation. This radiation can be partly stopped by some gases present in the atmosphere, the greenhouse gases, and returned to the Earth whose surface warms up accordingly.

Carbon dioxide is not the only greenhouse gas. Other gases such as methane, nitrous oxide and ozone produced by human activity contribute to the thermal imbalance of the Earth.

Agricultural activities are also responsible for emissions of greenhouse gases: in addition to CO_2 emissions due to the use of fossil energies, they also produce methane (from cattle and paddy fields) and nitrous oxide. Some industrial gases also participate in global warming, especially halogenated gases such as carbon tetrafluoride and sulphur hexafluoride.

The contribution of these greenhouse gases depends on their global warming potential (GWP), which measures the absorption of infrared radiation emitted back into the atmosphere. Carbon dioxide, with a GWP of 1, acts as reference. Methane has a GWP of 22, nitrous oxide 310 and sulphur hexafluoride 23 900!

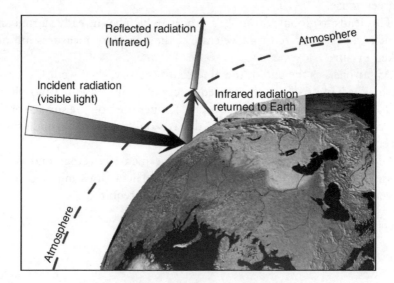

Figure 3.1   Greenhouse mechanism

The gas concentration and average atmospheric lifetime must also be taken into account. The effect of methane, with a lifetime of about 12 years, is therefore less damaging than one might expect in view of its GWP. We must nevertheless bear in mind that the mechanisms involved in atmospheric circulation of these gases and their exchanges with the soil and the oceans are complex. This leads to uncertainties regarding the long-term behaviour of CO_2 in the atmosphere and on the time required to return to equilibrium, in case of disturbance. The return to equilibrium occurs initially through absorption over several centuries of most of the CO_2 in the atmosphere by the oceans and the biosphere, followed by a very slow mineralisation reaction during which CO_2 is incorporated into carbonated rock, over a timescale of up to 100 000 years. This timescale extends well beyond our current forecast capabilities, especially since a new ice age is predicted to occur in 10 000 years. Whatever the case, spontaneous return to equilibrium can only be very slow.

Water vapour is also a greenhouse gas, but it condenses as clouds and does not accumulate in the atmosphere.

Of all the greenhouse gases, CO_2 is the one which contributes most to the greenhouse effect related to human activities, the anthropogenic sources, due to the considerable quantities emitted. In addition, since emissions of this gas are also increasing steadily, it plays a major role in the future risks of climate change.

The impact of greenhouse gas emissions on climate change

We have only recently become aware of the risks involved in climate change. In 1988, a decisive step was taken with the creation of the Intergovernmental Panel on Climate Change (IPCC), on the joint initiative of the World Meteorological Organisation (WMO) and the United Nations Environment Programme (UNEP).

The first IPCC report led in 1992 to the adoption of the 'climate' convention, currently ratified by 189 countries. The second report was published in 1996, before the signing of the Kyoto Protocol. The third report, published in 2001, led to the adoption of control measures and sanctions designed to ensure effective application of the Kyoto Protocol [19]. The fourth report, published in 2007, confirms and improves the main conclusions of the previous report.

Over 2000 scientists from 154 countries participate in the IPCC studies. Wishing to reach a consensus of opinion, the IPCC has remained prudent and it is therefore quite possible that the consequences of climate change will turn out to be more dramatic than those announced.

There is more and more evidence of the correlation between an increase in CO_2 emissions and an increase in the average temperature observed over the entire planet. Since 1860, the start of the industrial era, the average temperature on the surface of the Earth has increased by 0.8 °C.

Temperature fluctuations have already occurred in the past. Long-term climatic cycles are due to variations in the terrestrial orbit around the sun. This is the case in particular of the glacial–interglacial cycles, which have a frequency of about 100 000 years. We are currently experiencing an interglacial period and the trend towards a slight drop in average temperature, observed over the first eight hundred years of the past millennium, indicates a drift into a new glacial age in 10 000 years time.

The sudden rise in average temperature since the start of the industrial era seems to be abnormal compared with past trends, both in view of its relative amplitude and the speed of the change, on a geological timescale. Other phenomena of shorter frequency also occur, especially variations in solar activity.

This activity, which has increased over the past centuries, could partly explain the current warming, but certainly not the amplitude of the variation observed, nor the fact that since 1980 global warming has accelerated while solar activity has remained stable.

Some scientists, although very few, still express doubts as to the relation between climate change and greenhouse gas emissions related to human activities. The most vehement sceptics are in the USA, where the positions taken on climate change remain a political stake. Since it is becoming more and more difficult to deny the reality of climate change, the sceptics are now trying to denounce the way it is presented which, in their opinion, is excessively alarmist. This is the stance adopted by Richard S. Lindzen, a professor at MIT, who is the most frequently quoted advocate of this minority group [20].

The trend over the last few years has strengthened the conviction of most scientists that the rise in the average temperature observed is indeed correlated with human activities and, in particular, with the rise in atmospheric CO_2 content, which has changed from 270 ppm[1] around 1850 to 380 ppm in 2005. Figure 3.2 shows the trend in atmospheric CO_2 content.

This content has fluctuated over the last 400 000 years, just like temperature, whose variations are correlated with the atmospheric CO_2 content. The current variation nevertheless lies outside the fluctuation

---

[1] Parts per million.

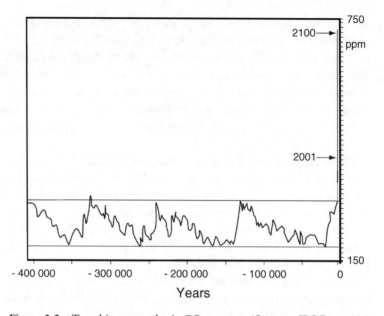

Figure 3.2 Trend in atmospheric CO_2 content (*Sources*: IPCC reports)

interval observed during previous geological eras. If the trend continues, the atmospheric CO_2 content could double by the end of the century, reaching a value of 750 ppm, and possibly exceeding 1000 ppm, causing disastrous climate transformations.

According to the models produced, the average temperature increase could reach 2–4 °C by the end of the century, possibly even 6 °C [9–12], while an increase of more than 2 °C is considered unbearable.

Symptoms of climate change

The last decades have witnessed various symptoms of climate change:

- *Generalised retreat of glaciers throughout the world.* The retreat is generalised in Europe and also in North and South America and Africa. The legendary snows of Kilimanjaro are now melting. The glaciers in the Andes and in the Rocky Mountains are also retreating.

  Rapid progress affects the Arctic ice. The ice pack has receded considerably and the thickness of the polar ice cap has shrunk 40% in about thirty years, from an average of 3.1 m to 1.8 m. By the end of the century, Arctic ice could melt completely in summer.

- *Regular rise in sea level*. The melting ice causes the sea level to rise: over the last twelve years there has been an increase of 3 mm per year, compared with 2 mm per year during the previous decades. The higher sea level is also due to heat expansion, which would account for about half of the rise observed. If this rise continues, numerous coastal regions could be flooded in the future, for example Bangladesh and the Ganges delta, the Maldives as well as a large part of Florida and the Netherlands.
- *Multiplication of heat waves and droughts*. Repeated heat waves and droughts have been observed over the last few years, especially in Europe. Although these episodes cannot be linked directly to climate change, their multiplication gives a further indication of the effects of climate change, to be added to the numerous other indications observed.

  During the 2003 heat wave, 20 000 deaths were recorded in Europe, including 13 000 in France.

  Eleven out of the last twelve years rank amongst the top twelve hottest years since accurate records have been available (since 1850). Various parts of the world, including Southern Europe, the Sahel and Australia, are suffering from serious drought, with increased risks of desertification. Heat waves and droughts lead to more forest fires, as we saw in particular in the summer of 2007, especially in Greece.
- *Cyclones and torrential rain*. Water evaporation caused by the higher temperature, responsible for the droughts observed, raises the air humidity which is responsible in turn for torrential rain in other parts of the world. Devastating floods have affected Europe and Asia, with greater rain fall related to the monsoon.

  A number of extremely violent cyclones, such as Katrina, have struck the world over the last ten years, although at this stage they cannot be attributed to climate change.
- *Impact on the biosphere*. Numerous effects of climate change on the biosphere can already be observed. Coral bleaching, which has been put down to the increase in water temperature, is one of the most visible consequences.

  There are numerous reasons accounting for the pronounced loss of biodiversity in the world but, for some species, the specific effect of climate change has been demonstrated. The consequences of the disappearance of these species gradually affect the entire ecosystem in which they live.

Higher temperatures also result in displacement of ecosystems. A 3 °C temperature increase, for example, corresponds to a shift of 300–400 km in latitude. This may prove fatal for an existing ecosystem, for example by favouring the propagation of a parasite not previously present.

The relation between the various phenomena observed and global climate change must be analysed with a certain degree of caution, given the complexity and variability of meteorological phenomena. The fact that they occur simultaneously is nevertheless the symptom of a dramatic change. In addition to effects visible locally, we also observe profound changes affecting the entire planet.

Positive retroaction and tipping points

The so-called 'positive retroaction' mechanisms are likely to make climate change even worse. Positive retroaction corresponds to higher CO_2 emissions, with anthropogenic emissions releasing greenhouse gases trapped in the soil, the ocean, the earth and the biosphere.

Large quantities of methane trapped in the *permafrost*[2] could therefore be released due to a temperature increase. The methane is trapped in crystals formed in the presence of water, the hydrates. Once the temperature is high enough to cause decomposition of the hydrates, methane is quickly released. Since methane has a relatively high GWP, there is a considerable risk for the planet. The Siberian permafrost has started to thaw, suggesting that this change could occur in the relatively short term.

Maritime streams such as the Gulf Stream could be modified, with consequences for the climate of the coastal regions. Climate change could have also an impact upon interaction phenomena between ocean water layers and winds, such as those which produce El Niño. El Niño is a strong maritime stream, which occurs periodically in the Pacific Ocean. It is provoked by dominant winds, which induce an oscillating mechanism in the atmosphere coupled with a displacement of surface ocean layers. This oscillating mechanism results in modifications of climatic conditions and of rain occurrence.

Disastrous effects have been observed as a consequence of El Niño, such as droughts and famines, forest fires and floods[3], which illustrate the possible consequences of such a climatic disorder.

---

[2] Permanently frozen ground of the Arctic regions.

[3] The disastrous episode of massive forest fires which struck Indonesia in 1997 can be related to El Niño.

Other positive retroaction phenomena are likely to increase the imbalance: thus, disappearance of the reflective effect of pack ice with respect to solar radiation, the so-called 'albedo' effect, will contribute to increasing the quantity of heat absorbed by the ground and therefore to heating the surface.

The temperature increase could also affect the tropical forests, which currently capture some of the CO_2 in the atmosphere. If these forests should disappear, they would release the carbon they have stored until now. These risks concern in particular the subtropical forests, whose disappearance could accelerate if the temperature reaches a certain threshold. A reduction in their area, by modifying the rainfall, would reduce the quantity of water received by the remaining forest, causing a chain reaction. Repeated heat waves also promote outbreaks of forest fires. The Amazonian forest could therefore disappear in a relatively short time.

Lastly, a major risk identified by the IPCC in its last report concerns a reduction, caused by a rise in temperature, of the ability of the oceans to act as a carbon sink, thereby increasing the atmospheric CO_2 content.

All these examples demonstrate that the effects of climate change could increase suddenly above a certain point which could be reached in the relatively short term. Survival of all ecosystems on the surface of the Earth is concerned with such 'tipping points'.

The dramatic consequences of a temperature elevation are increasing rapidly. For a 1 °C elevation, it is estimated that acidification of the oceans would cause serious damage to the coral reefs and to the ecosystems of the Arctic region.

For a 1.5 °C elevation, we would observe irreversible thawing of the polar ice cap in Greenland. Between 2 °C and 3 °C, numerous ecosystems, especially the Amazonian ecosystem, would be threatened. At 3 °C, we would be faced with conditions leading to massive extinction of numerous plant and animal species, under conditions similar to those which led to the major extinctions which occurred during the geological ages [21].

It therefore seems vital not to exceed a temperature elevation of 2 °C. According to the models produced, to achieve this it is essential to limit the atmospheric CO_2 content which will be reached in 2050 to a value of between 400 ppm and 450 ppm.

The vulnerability of each country varies according to its geographic position and its ability to adapt. The northern industrialised countries are less vulnerable than the developing countries in the southern hemisphere, in view of their latitudes and also their ability to invest to adapt to climate change. Emerging countries like India and China are highly exposed to the risk of climate change, which could hinder their development. A very large proportion of the world population is therefore seriously concerned [22].

Should we consider that there will be winners and losers, and that the countries in the north (or the extreme south) will benefit from climate change? Such an attitude would be hazardous and risky. Certain countries could partially benefit from the situation, provided that the climate change remains moderate. If the change was to intensify, they would themselves suffer from the dramatic consequences, and the tensions resulting from the distress of the worst hit countries would escalate out of control. Famine, water shortage, floods and the increased risks of epidemics would inevitably lead to massive migrations and conflicts which would be all the more fierce for being quite literally a fight for survival. Despite the remaining uncertainties it is therefore urgent to combat global warming, which would threaten our entire planet, with every possible means.

According to the Stern report, if no appropriate measures are taken immediately, the financial impact of climate change would amount to over 5500 billion Euros [13]. It reveals that inaction may turn out to be far more expensive than the measures which must be taken immediately to prevent any worsening of the phenomena related to climate change.

A vigorous action plan is required. The Kyoto Protocol raised public awareness regarding the problem of climate change, but much more radical measures must be set up in the future.

Kyoto protocol and post-Kyoto

In 1997, the 188 signatory countries met at Kyoto in order to agree on a number of restricting commitments.

Amongst the signatory countries, the 38 'Annex 1' industrialised countries committed to reducing their CO_2 emissions by 5.2% from 1990 levels between 2008 and 2012.

The Kyoto Protocol only came into effect in 2005, after its ratification by Russia. To date, 172 countries have ratified the Protocol, excluding the USA which refused to sign.

The European Union as a whole made a commitment to reduce its emissions by 8%. The efforts to be made by each country varied, with France simply having to stabilise its emissions.

To reach these objectives, each country may not only use a complete range of internal measures (energy taxation, energy saving incentives, etc.) but also resort to three 'flexibility' mechanisms:

- The Joint Implementation (JI) mechanism establishes the possibility for an Annex 1 country to obtain CO_2 emission credits by investing in

a greenhouse gas (GHG) emission reduction project from another Kyoto Protocol signatory country.
- The Clean Development Mechanism (CDM) is based on the same principle as that of the JI mechanism, but in this case, the investments are made in countries such as China, India and Brazil which are non-Annex 1 countries.
- The CO_2 Emissions Trading Scheme (ETS) allows for the possibility of exchanging the CO_2 quotas assigned to companies. At the end of each period, a company whose actual CO_2 emissions are greater than its allowances can purchase the missing allowances from the market. This is only possible if, similarly, other companies emit less than their allocation of allowances and are in a position to sell their surplus allowances.

Of the major industrialised countries, only the USA has not signed the Kyoto Protocol, since Australia has just signed. A certain number of American states, including California, have nevertheless adopted a very similar system.

At the same time, it is quite obvious that some parties will not respect their Kyoto commitments.

On 1 January 2005, the European Union set up the European Emissions Trading Scheme, provided for under the terms of the Kyoto Protocol.

Companies which make investments to reduce their CO_2 emissions may, if their performance is better than the objective set by the government, sell emission allowances and thereby pay back their investments. Stock exchange mechanisms apply in this case: the value of a tonne of CO_2 varies according to supply and demand and according to the volumes traded. This system will be restricted to Europe (some 12 000 industrial sites are concerned) for 3 years before being extended internationally in 2008, thereby concerning exchanges not only between companies but also between countries. The aim of this system, by combining government intervention (to set reduction objectives and check compliance with commitments) and market mechanisms, is to promote collective and global control of emissions. The idea is to optimise investment levels, which will primarily concern the most effective actions in terms of reducing CO_2 emissions, through market regulation either in Europe or in the developing countries by implementing clean development mechanisms.

After a sharp increase, the value of a tonne of CO_2 dropped to a very low level due to a surplus of allowances for the first application period. A readjustment was made in the second application period of the system (2008–2012), with rates in the region of € 20 per tonne of CO_2.

The Kyoto Protocol objectives seem relatively modest compared with the requirements. The Kyoto Protocol nevertheless offers the major advantage of having initiated a process and having set up mechanisms aimed at reducing CO_2 emissions.

The question of the post-Kyoto period now arises: what actions must be taken after 2012? A new step must be defined to move towards the targeted objective.

The European Union has demonstrated its determination to make further progress, by setting itself the objective of a 20% cut in greenhouse gas emissions by 2020. The actual means required to reach this objective still remain to be defined.

The economic mechanisms to limit greenhouse gas emissions will only be truly effective if they are generalised. The introduction of emission allowances or a carbon tax has major impacts on the economy [32, 33]. It is important to avoid the negative impact of distortion of the competition between countries which would not apply the same rules. This could result in delocalisation of the most CO_2-polluting industries to countries whose environmental regulations are most lax.

Consequently, there is an urgent need to reach an international agreement covering the post-Kyoto period, with precise figures on the reduction of CO_2 emissions which would be accepted and applied by as many countries as possible and especially by the most polluting countries.

The factor 4 problem

Various scenarios concerning climate change and CO_2 emissions have been produced, especially by the IPCC.

In the reference trend scenario shown in Figure 3.3, the emissions of carbon as CO_2 related to energy consumption change from 7 to 14 Gt of carbon per year between the current period and 2050, i.e. from 26.6 to 52 Gt of CO_2 (trend A). Such a trend would produce a level of CO_2 in the atmosphere which is totally unacceptable in terms of impact on the climate.

We must therefore consider an alternative scenario on CO_2 emission trends which would be compatible with the aim of limiting the average temperature increase to 2 °C compared with the pre-industrial situation.

Scenarios of this type have been drawn up in the European Union in the context of the *Greenhouse Gas Reduction Pathways* (GRP) study.

According to the two objective scenarios, the concentration levels of the six greenhouse gases would stabilise respectively at 550 ppm and 650 ppm

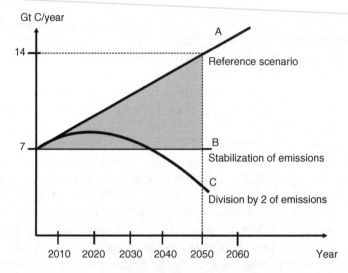

Figure 3.3   Trend of CO_2 emissions in tonnes of carbon per year

(450 ppm and 550 ppm for CO_2 alone). In the first case, the temperature increase estimated on the basis of the IPCC results would be 1.6 °C and in the second case 2.5 °C [34].

If the CO_2 emissions were stabilised now (trend B), the CO_2 content in 2050 would be about 500 ppm. In practice, the possibility of stopping the rise in emissions immediately would be unrealistic, and they are likely to continue rising until 2020, even if we manage to slow down the rate of increase. Emissions will then have to decrease to reach a level close to half the current level by 2050, in order to respect the limit of an average temperature rise of 2 °C (trend C – scenario 450 ppm for CO_2 alone).

If the long-term objective is to express the level of CO_2 emissions as a figure per inhabitant in the world, which would seem fair, the emissions from the industrialised countries must be reduced by a factor of about 4 by 2050, in order to allow the emissions from the Third World countries and the developing countries to increase. France has assigned itself this objective [35].

The current trend

Compared with these objective scenarios, the actual trend observed is worrying. Current CO_2 emissions are in the region of 27 billion tonnes per year due to the consumption of fossil fuels. Electricity production (39%), industry (22%) and transport (23%) represent, at world level, the main sources of CO_2 emissions related to human activities.

The current CO_2 emission growth rate, far from declining, is tending on the contrary to accelerate due in particular to the rapid development of China. China, which relies heavily on coal, is becoming the main CO_2 emitter in the world, ahead of the USA. In 2006, China's CO_2 emissions were 8% higher than those of the USA, while in 2005 they were 2% less and in 2000, 28% less.

At world level, between 2000 and 2004, the total CO_2 emissions increased at a rate of 3.1% per annum, as compared with 1.1% during the previous years.

If the current trend continues, by 2050 CO_2 emissions would increase, not by 100%, as indicated in the reference scenario mentioned previously, but by 130% with an annual growth rate of 2% and up to more than 250% with an annual growth rate of 3%. The current trend is therefore moving in the direction exactly opposite to that required. To avoid catastrophic and irreversible climate change, we must rapidly reverse this trend by implementing, as a matter of urgency, a large-scale action plan designed to limit CO_2 emissions.

Urgency of the actions to be undertaken

The urgency of the measures to be taken is sometimes discussed. Two types of argument are put forward to delay the decisions:

- A first argument is to consider that the future remains uncertain. The degree of uncertainty is considered as an argument for biding time, with the hope that in the future we will be in a better position to assess the situation and optimise the strategy to be applied.
- A second argument is based on the idea that new technologies will emerge during the next few years, placing us in a better position to tackle the problem. This opinion, which reflects the stance taken by the current American administration, relies on the development of new technologies to provide the right solutions in time.

In addition to these arguments, we are faced with the psychological difficulty of tackling a future catastrophe, which has not yet happened even if it is imminent. As J.-P. Dupuy[4] wrote: 'knowing is not believing' and it is difficult for us to imagine a future catastrophe and even more

---

[4] A French scientist and philosophe.

difficult to take action accordingly [36]. These arguments and this psychological context merit close analysis.

Analysis of the potential impact of climate change indicates that it is essential to act immediately, by launching a decisive action plan designed to reduce CO_2 emissions:

- There is increasingly convincing evidence demonstrating the risk of major catastrophes if a suitable response is not found. The observations and the forecasts of the models available all converge to the same conclusion.
- If the actions required to reduce CO_2 emissions are delayed, it will prove increasingly difficult to respect the limits which must not be exceeded to avoid catastrophic effects. The longer we continue behaving as we do now, in other words increasing the quantities emitted every year, the more drastic the future reduction will have to be; we will reach a point of no return when it will be quite impossible to reach the targeted objective.
- The more we deviate from the current equilibrium, the more the uncertainties concerning the maximum amplitude of the climatic differences increase, with a rapidly growing risk of totally uncontrolled phenomena. In addition, positive retroaction effects, such as the release of methane trapped as hydrates or the disappearance of a sea current may occur above a certain threshold, with sudden consequences.

The urgency of the measures to be taken is therefore evident. Given the contribution of fossil energies to CO_2 emissions, it is vital to initiate without delay the energy transition which will allow us to prevent a major energy crisis as well as an irreversible climatic catastrophe. Strong action by the Public Authorities is therefore necessary, firmly backed by public opinion in the various countries.

4

The Essential Energy Transition

The need for change

In view of the threats poised over our economy and our planet, due to excessive consumption of fossil fuels, a transition to a new energy system is essential.

Such transitions have already taken place in the past, when coal replaced wind and waterpower, and then when oil took over from coal with the advent of the automotive and aviation era.

Now, however, we are faced with a quite different situation. The previous transitions were the result of technological progress, which progressively modified the economic system, without this transformation having been anticipated.

In contrast, the energy transition we are now facing is imposed upon us, but there are no obvious substitution options.

Its implementation is made difficult, however, due to the very high level of worldwide energy demand and the magnitude of the threats to the environment. For a number of economic and technical reasons which will be explained below, the fossil energies cannot be replaced rapidly and massively by alternative energies (renewable, nuclear).

Although the energy transition promises to be long and difficult, it must be undertaken without delay. Solutions specifically adapted to this transition period will have to be proposed.

Energy and Climate: How to achieve a successful energy transition   Alexandre Rojey
Copyright © 2009 Society of Chemical Industry

Threats to be removed

To cope with the numerous threats facing us (progressive depletion of resources, unequal access to energy, tensions over supplies and damage to the environment), the necessary solutions must guarantee:

- Long-term availability of the energy required for development of the planet, without excluding the Third World countries.
- Security of supplies, to protect the importing countries against sudden crises.
- Availability of the means designed to avoid the catastrophic effects of a major climate change.

These threats must be considered in a broader context. The risk of depletion does not concern energy alone, but all natural resources (water, food resources, raw materials, etc.). The notion of scarcity, which plays a major role in the economy, requires new forms of management and governance [23].

Pollution and global warming could make the planet unfit to support life. Pollution of the natural environment coupled with excessive use of pesticides and dumping of waste worsens the situation, especially with regard to access to water.

Overexploitation of water resources for irrigation or consumption in towns depletes the water tables, dries out rivers and causes numerous lakes to disappear [97]. Global warming also jeopardises water resources in many regions.

The concept of an **ecological footprint** formalises the problem of access to the Earth's resources. The geological footprint represents the area of the Earth needed to regenerate the resources a human population consumes and to absorb and render harmless the corresponding waste. It is greater than 5 ha per capita in the richest countries and less than 1 ha in the poorest. In 2003, it was 5.6 ha in France and 9.6 ha in the USA. The ecological footprint of the world population is rising steadily, and it is estimated that it has changed from slightly over 4 billion ha in 1963, to nearly 14 billion ha in 2003. In contrast, the maximum footprint per capita that can be supported in terms of natural resources is consistently decreasing, changing from 2.9 ha in 1970 to 1.8 ha in 2003. If every inhabitant on the planet was to consume as much as a current inhabitant of the USA, the equivalent of 5.3 planets would be necessary to support humanity, clearly demonstrating that this option is not viable [9].

Pressure on the environment is likely to increase further, with major consequences for the ecosystems on Earth. Biodiversity is especially vulnerable.

Unless immediate action is taken to remedy the current practice, this situation could eventually become disastrous, causing an ecological catastrophe [16], a scenario already predicted by some authors [14, 15].

Acting from a perspective of sustainable development

The various threats discussed demonstrate the need to act from a perspective of sustainable development. Sustainable development was defined in the Brundtland report presented to the United Nations in 1987 as a 'development that meets the needs of the present without compromising the ability of future generations to meet their own needs.' [27].

The world's resources are finite and must be managed accordingly. Similarly, our environment has only a limited capacity to react to attacks of pollutant discharges and must be protected. In 1972, the Meadows report commissioned by the Club of Rome raised the question of the compatibility between economic growth and maintenance of natural equilibrium [28].

The aim of sustainable development is precisely to reconcile these two requirements and find a compromise between a position which favours growth, irrespective of the consequences on the environment, and one which consists of accepting no change in the environment, regardless of the economic implications. The analysis of the means necessary to achieve this aim nevertheless varies substantially depending on the authors and several models of 'weak' and 'strong' sustainability have been put forward. These models diverge on the possibility of introducing alternative resources to compensate for the natural resources consumed [29]. The advocates of 'strong' sustainability reject any changes to the state of the planet resulting from human activity and therefore any irreversible consumption of natural resources.

The very concept of sustainable development is questioned by those who consider that it combines two incompatible notions and urge economic growth as the sole means of preserving our environment.

It seems difficult to deny the developing countries the right to better welfare, which cannot be obtained without energy consumption. It is nevertheless possible to conceive new ways of living offering better welfare and quality of life, while moderating energy consumption.

Lower pollution and the resulting positive consequences on health represent an essential factor to improve the quality of life.

Energy choices cannot be dissociated from the major problems which will affect humanity in the coming years: demographic control, consumption modes and standard of living, education, health, food and water requirements. Faced with the various threats, a global transition must be successfully implemented to drive back under-development and preserve the planet for future generations.

The current energy model of the richest countries is neither durable nor exportable. To prevent tensions which would be likely to degenerate into violent conflicts, we must reconsider the way energy is used, taking into account improvement of the well-being of the entire population.

Initiative at a global scale will be necessary to meet the planet's requirements. The Kyoto Protocol represents a first attempt which, although incomplete, has the merit of triggering the adaptation required to tackle the risks of climate change.

The very notion of transition may seem to be in contradiction with the concept of durable development. The solutions to be implemented during the transition period are not truly 'durable', since they aim at simplifying the creation of a different system using measures including some which will be temporary. The transition will produce a new energy system that will be more 'durable' than the current system, but will probably be neither total nor permanent.

It is therefore more accurate to speak of 'sustainable' development. In a changing world, no system can continue indefinitely. It is essential, however, to make our economic development compatible with the survival of our planet.

The need for global regulation and governance

The risks of climate change raise a problem that humanity has never had to face before.

The situation created by massive CO_2 emissions, unlike the problems of local pollution already encountered, is unprecedented. It is exceptionally serious and affects the entire planet. For the time being, however, its consequences are only felt very slightly by the human population.

There is little chance that the measures required will follow naturally, based on current market mechanisms alone. These measures have a relatively high cost and it seems unlikely that an economic player will accept this cost, without some sort of statutory or incentive mechanism.

The cost related to protection of the environment must be included in concrete terms in the production or consumption process. This means that the polluter (whether an individual or a company), must pay a cost equivalent to that required to remedy the pollution caused.

Until now, the atmosphere has been considered as free common property. As a result, considerable quantities of greenhouse gas have been discharged until quite recently with no penalty. To limit these discharges, it is essential to create a mechanism, for a group of countries and eventually at global level, aimed at capping CO_2 emissions or limiting these emissions by setting up a taxation system.

Generalisation of a tax on fossil energies, related directly to CO_2 emissions, is frequently mentioned [32]. This type of tax is neither easy to establish nor popular. It nevertheless remains clear that a statutory mechanism is required to levy the additional costs incurred in order to protect the environment.

The 'compensation' principle is more and more frequently put forward. Under this principle, the 'carbon neutrality' of an activity responsible for CO_2 emissions is guaranteed by financing, through an approved organisation, a compensatory action leading to an equivalent reduction of CO_2 emissions somewhere else on the planet. It is therefore possible to 'compensate' for the CO_2 emissions caused by an air trip.

The principle of this mechanism is not fundamentally different from that implemented in the emissions trading scheme, described in Chapter 3. Although currently applied on a voluntary basis only, it offers the advantage of involving all sectors, including those outside the current emissions trading scheme, and especially transport.

The European Union has an important role to play in this area. By setting up governance across the 27 member countries, it can demonstrate that this system is feasible. Provided that other parts of the world (North America, Asia-Pacific) set up similar systems, its extension to every country of the world could become effective through a scheme managed by the United Nations.

Even though it is difficult to reach an international agreement, ratification of the Kyoto Protocol represents a highly encouraging step. In the past, similar international agreements have been signed, especially to ban the chlorofluorocarbon (CFC) compounds used as propellants in aerosols which threatened the ozone layer (see Chapter 2). In 1987, the Montreal Protocol opened the way to halting the emission of substances depleting the ozone layer. Since then, the statutory measures have been reinforced and, in 1996, the start of a decrease in atmospheric CFC content was observed [7].

Energy alternatives

Faced with the major risks which have been mentioned concerning energy supplies and climate change, it is essential to adopt more durable solutions which do not suffer from the disadvantages of fossil energies in terms of greenhouse gas emissions and depletion of resources.

The objective is therefore to reduce as quickly as possible the proportion of fossil energies in the global primary energy supply, while supporting the vital needs of the world population.

We must therefore look towards alternative energy sources, namely nuclear and renewable. In principle, the renewable energy resources are inexhaustible on a human scale.

With the future technologies, the nuclear sector will be capable of producing energy for a very long period of time, despite the issue of uranium availability for the nuclear power plants based on current technology (see Chapter 6).

Presently, however, these alternative energies represent only a very small fraction of the world's total primary energy supply (less than 20%). Moreover, this share is only growing very slowly.

The alternative solutions (nuclear and renewable) pose problems in terms of technological maturity and economic profitability in the case of massive development of the renewable energies and, as regards nuclear energy, there is also the question of safety and social acceptability.

Until new technical breakthroughs allowing broader distribution of these alternative solutions become available, we must implement all possible means to make the transition without suffering a major crisis.

A long transition

It will take a long transition period to reverse the respective shares of fossil and non-fossil energies, starting from the current system which is 80% based on fossil energies.

Figure 4.1 illustrates the predicted trends in non-fossil energies for the period 2000–2100. The general trend expected is an S-shaped curve, changing slowly at first then accelerating up to a point of inflexion before slowing down at the end of the transition.

To reverse the respective shares of fossil and non-fossil energies by the end of the century, the point of inflexion must occur at 2050, with a share of non-fossil energies of about 50%. Most analyses conducted for periods up to 2050 predict a significantly lower share of the non-fossil energies,

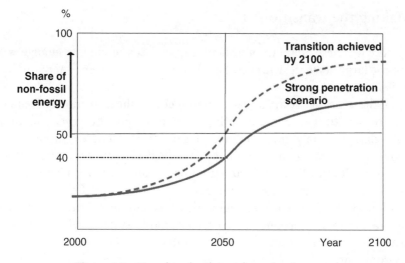

Figure 4.1   Trend in the share of non-fossil energy

within a range not exceeding 30–40%, even given a favourable scenario, with assumptions aimed at favouring the penetration of non-fossil energies [3, 17–19].

The global transition, resulting in significant replacement of fossil energies by non-fossil energies, will therefore last a considerable period of time. Even though it has already started, it will probably not be completed before 2100.

This duration is due to the time required to develop alternative solutions and the intrinsic inertia of the energy sector. The investments made for energy production facilities such as power stations are repaid over periods extending up to thirty or forty years. In the automotive industry, the time required to design a new type of vehicle and setting up the production lines is also very long. Introduction of new equipment is faster in the emerging countries, but the top priority of these countries is to ensure their own development rather than reduce their CO_2 emissions.

Also, in a certain number of applications, no immediate alternatives are available. This is the case, in particular, for the transport sector, which is almost exclusively dependent on oil.

Even when these alternatives are available, their distribution takes a certain amount of time since equipment is only renewed relatively slowly: this is true of the housing sector, in particular.

Given the urgent need to find solutions to the various threats and in particular to limit climate change, it is obvious that we cannot wait for completion of a long drawn-out transition and concrete measures must be taken quickly to face the problems posed.

Making the transition

Faced with these adaptation difficulties, the risks of climate change will call for large-scale measures to be taken rapidly in order to reduce CO_2 emissions.

It is therefore necessary and urgent **to speed up the transition movement** with respect to the trend currently accepted by most observers. In addition, changing only the structure of the primary energy supply (also known as the 'energy mix') will not be sufficient to ensure a satisfactory transition. It must be combined with other means, described below, such as improved energy efficiency and geological storage of CO_2.

A long-term change is also required, which comprises the steps necessary to eventually produce a more sustainable system, while avoiding situations which would be unacceptable for the economy or the environment.

This will involve considerable changes. The trend observed in the years to come will be decisive for the future. Depending on the decisions taken, the energy transition could succeed and tend towards a sustainable equilibrium or, on the contrary, fail and lead to a crisis accompanied by major conflicts [24].

Numerous examples from the past show that societies which were unable to adapt to a change in their environment, especially depletion of resources, have collapsed suddenly. Some authors already consider that our society could suffer the same fate [25, 26].

This prediction is all the more worrying since, due to globalisation of the economy, such a collapse would not concern a single country or region, but the entire planet.

The step to be taken is therefore extremely critical, since events could accelerate and the risk of catastrophe is not excluded. An action plan must be implemented as a matter of urgency; the effort to be made will extend over a long period of time, until the threats have been removed. The transition must be initiated in the richest countries, which will be followed, with an inevitable delay, by the developing countries.

The need to innovate

A great deal of creativity and innovation will be required at all levels to implement the right solutions on a large scale over the coming years. All the solutions currently available must be implemented without delay, while at the same time developing those for tomorrow.

Studies must be conducted on new systems for production, transformation and use of energy in order to consume less energy and emit less CO_2 [30]. Carbon capture and storage or recycling techniques must also be developed. Higher efficiencies and greater use of renewable energies will necessitate new energy storage systems and new energy vectors. This will be impossible to achieve without a major research and innovation effort.

The creative effort required does not simply concern technical innovation but also ways of life and behaviour. We will have to take a fresh look at habitat and mobility to better adapt them to the new constraints, introduce new consumption modes and modify working methods, while preserving and, whenever possible, improving the quality of life.

Appropriate regulation systems must also be defined and applied on a global scale to protect the environment. Creation within the European Union of an emissions trading scheme represents an interesting example of an economic mechanism which favours a reduction in CO_2 emissions.

Without a strong political commitment from the main emitting countries, it will not be possible by 2050 to halve the global CO_2 emissions compared with the current level, or even reduce them to this level.

The creation of programmes involving different countries and international groups of experts such as the IPCC (Intergovernmental Panel on Climate Change) should help develop the necessary measures on an international scale.

These various actions are complementary. They imply collective awareness regarding the urgency of the actions to be undertaken and demand creativity in all areas. The exceptional tools currently available to observe the planet (in particular satellites) and the development of communication systems (internet, TV) are invaluable vectors to generate this awareness.

The risks involved in the solutions proposed must nevertheless be assessed. Some appear risky: this is the case, for example, with injection of CO_2 into the oceans and 'geo-engineering' solutions, which consist in reducing the transmission of solar radiation through the atmosphere by injecting sulphur particles to create aerosols which increase the atmospheric albedo.

Means of action

To support the global energy demand essential for development of the planet while avoiding a climatic catastrophe, action must be taken along two axes, as illustrated in Figure 4.2.

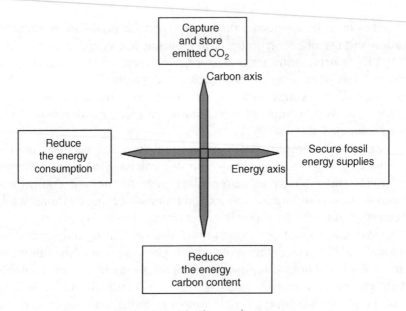

Figure 4.2   The four action points

• Energy supply and demand axis

The future balance between energy supply and demand must be obtained first by reducing the demand. Energy must be used more effectively, to reduce consumption without jeopardising development. At the same time, a sufficient supply must be provided by diversifying the energy supply sources.

• Carbon balance axis

To limit the risks related to climate change we must control CO_2 emissions, by reducing energy consumption, and also the carbon content of the energy *mix*. Unfortunately, the anticipated comeback of coal will contribute to increasing the carbon content. To reach the CO_2 emission reduction objectives, we must also find other solutions and, in particular, resort to capture and storage of at least a large proportion of the CO_2 emitted.

Although not easy to establish, a balance must be found between economic development (supply–demand axis) and essential protection of the environment (carbon balance axis).

To guarantee the security of energy supplies while preserving the environment, an ambitious action plan matching the stakes involved must be initiated without delay, based on the following four action points:

- *Reducing energy consumption* while continuing economic development, especially in the poorest countries, and protecting the environment, is the top priority. This is undeniably the best way of reducing energy dependency on hydrocarbon imports, while at the same time limiting CO_2 emissions. It also represents a way of promoting more harmonious relations between economy and society.
- *Reducing the energy carbon content* to decrease CO_2 emissions per unit of energy produced favours energy sources with reduced carbon content (nuclear and renewable) and offers the opportunity of examining how to globally reduce the CO_2 emissions of the energy system.
- *Securing fossil energy supplies* remains vital throughout the transition period. As we have seen, this involves first limiting the consumption of fossil energies. To avoid a major crisis in oil supplies during the future transition period, however, we must also develop new ways aimed at pushing back the production limits through new discoveries, better exploitation of reserves already identified and putting into production new resources, such as heavy and extra-heavy oils. To reduce the risks concerning future supplies, the source of the fossil energies used must be diversified. We must also avoid shifting the demand onto the worst CO_2 emitters, in particular coal.
- *Capturing and storing the CO_2 emitted* Since the previous measures might fail to reduce CO_2 emissions quickly enough, additional means must be implemented to achieve this: carbon sinks, capture and geological storage of CO_2, and carbon recycling.

These four action points are discussed in Chapters 5–8.

Technologies specifically adapted to the transition period must be applied to simplify the move to a new energy system. Priority will be given to those solutions that can directly replace the current solutions.

In the transport sector, liquid **biofuels**, which can replace the current fuels, can be implemented rapidly since there is no need for a radical transformation of vehicle design or the distribution system. We must nevertheless ensure that the production of biofuels respects the sustainability criteria (see Chapter 6).

Hybrid systems combining the use of a fossil fuel and zero carbon energy will become increasingly important. They allow the use of an intermittent source of energy production (solar, wind power) while benefiting from the advantages offered by fossil fuels in terms of production flexibility. The hybrid propulsion systems which can be recharged on the mains electricity supply, to be detailed in Chapter 6, are another example.

More generally, the energy transition requires the implementation of hybrid systems combining several forms of energy, different fuels and propulsion modes, in order to progressively increase the proportion of non-fossil energy in the energy mix.

Flexible systems, which can operate on different forms of energy, simplify the transition by allowing the user to opt for an alternative energy without having to make an irreversible choice from the outset. Flex-fuel vehicles (FFV) which can run on gasoline, ethanol (E85), or any mixture of the two, are one example. Vehicles equipped with dual fuel engines, that can either run on liquid oil fuel or on LPG or NGV, represent another example of transition technology.

The energy **storage** systems must become more efficient and more cost-effective. Fossil fuels store a large amount of energy in a limited volume and can therefore be stored very easily, especially in vehicles. This is not the case with most alternative energy sources which must be converted into electricity (nuclear, wind power, photovoltaic solar power, etc.). Greater recourse to the renewable energies due, for most of them (apart from ex-biomass and geothermal energy), to their intermittent nature, will imply a need for energy storage systems.

Increased deployment of **energy vectors** compatible with the use of different energy sources also represents one way of facilitating the transition. Greater reliance on electricity will therefore tend to favour diversification of energy sources. Eventually, the use of hydrogen as an energy vector may play a similar role and help extend the number of applications potentially concerned.

In the future, the energy system will rely on extended interconnected energy networks, offering more flexibility than the current electricity networks, and on efficient energy storage systems.

Decentralised electricity production systems could connect to networks like this, with the prospect of a progressive growth in renewable energies, in particular wind and photovoltaic power.

A global transition

The transition to be made is not related to energy alone, it must be much more global. The other transitions to be implemented concern in particular:

- *Demography*. The world population has exploded from 3 billion inhabitants in 1960 to 6.5 billion inhabitants in 2005. The growth rate which peaked at 2.1% between 1965 and 1970 had dropped back to a level of about 1% for the world population. This global trend covers major disparities, the fertility rate of the poor countries (2.9 children per woman) being roughly double that of the rich countries (1.6 children per woman). The population will reach 9.1 billion inhabitants in 2050, if the fertility rate continues to decline progressively.

If the demographic pressure on the environment is to diminish, we must ensure that this deceleration in demographic growth rate is kept under control and stabilise the level of the world population. After the current transition period during which the birth rate exceeds the death rate, we could return to a new equilibrium, with a fall in the birth rate, resulting in a stable world population [39].

- *Food production.* We must be able to feed the world population, while avoiding excessive use of energy and resources, via new agricultural production modes and also by developing novel consumption modes.

  Improving the efficiency of the food chain is of paramount importance to cope with the increasing consumption of proteins in the world. We must move towards breeding species which consume the lowest quantities of cereals, while striving to moderate the consumption of meat in the richest countries. In this respect, the rapid growth of dairy production in India and fish farming in China constitute significant achievements [97].

- *Water and raw material resources.* New raw material management modes must be found to minimise losses, based on recycling. It is essential to reduce pollution which continues to spoil vital resources, often irreversibly. The issue of water resources clearly represents a major concern for the future.

- *Development and education.* Success in the various transitions required, and primarily the demographic transition, will depend on progress in eradicating the most extreme forms of poverty and in the field of education.

- *Ways of life and mentalities.* This 'cultural' transition is essential, especially in the richest countries. These countries must manage to change their ways of life in order to dissociate creation of value and quality of life from unrestrained consumption of energy resources.

The energy transition must therefore be considered from the perspective of a global transition which concerns society as a whole through its ways of life and the way it manages resources.

The issues of energy, development and the environment are interlinked and require a common approach guided by true 'planetary ethics'.

5

Reducing Energy Consumption, while Protecting the Environment

From megawatts to negawatts

The first priority in order to ensure the energy transition in the best way is to reduce energy consumption. Reducing this consumption contributes to decreasing the dependence on energy supplies and to eliminating the corresponding CO_2 emissions, together with all other environmental impacts resulting from the production and use of energy.

Within this context, any investment in reducing energy consumption is most suitable for resolving all the problems previously mentioned. Each watt saved while keeping the same level of satisfaction for the final user represents what can be called a 'negawatt'[1]. A negawatt has a higher value than an additional watt, because it can help to provide the same use or service, while avoiding the negative impact upon the environment of a supplementary production of energy. Replacing 'megawatts' by the production of 'negawatts' is a most sensible choice, whenever it is possible.

The efforts in this area are very sensitive to the price of energy. The first and then the second oil crises led to the setting up of energy saving policies throughout the world. The subsequent counter oil crisis has resulted in a detrimental demotivation.

---

[1] The association Négawatts has been created in France for promoting this concept.

Energy and Climate: How to achieve a successful energy transition   Alexandre Rojey
Copyright © 2009 Society of Chemical Industry

Presently, the risks of climate change represent a strong supplementary reason for reinforcing energy conservation.

The actions which are required concern both consumption modes and the technologies to be implemented. They need to be set up at three levels:

– Avoiding the waste of energy.
– Adopting new organisation and consumption modes that are less energy intensive.
– Investing in energy conservation technologies.

Investments in the area of energy conservation are also an excellent choice in economic terms, as they represent protection against future increases in the price of energy.

The evolution of the energy intensity

Energy consumption increases with the Gross Internal Product (GIP)[2] once a certain level has been reached. The energy intensity, defined as the energy consumption related to the GIP, can be used as an indicator of the efficiency in the use of energy for ensuring living standards in a country and its evolution with time.

Large discrepancies are observed between the energy intensities of different countries in the world. Thus, the energy intensity of the USA is higher by 50% than the average energy intensity in the European Union. The differences result from a whole set of factors such as the latitude, which determines the needs for residential heating, or distances, which influence the need for transportation. They result also from different ways of life.

The diagram in Figure 5.1 shows the important progress that has already been accomplished and that remains possible [37]. Thus the US energy intensity has been divided by a factor close to 2 during the last thirty years. The energy intensity of France has been reduced by 35 % between 1973 and 2001.

In developing countries, the energy intensity initially grows during the first development phase and then decreases. Thus the energy intensity of China has initially reached high levels and has sharply decreased later on. The present level is slightly above 0.2 toe/1000 US $ GIP (0.23 toe/1000

---

[2] Total value of the internal production of goods and merchant services within a country during a year.

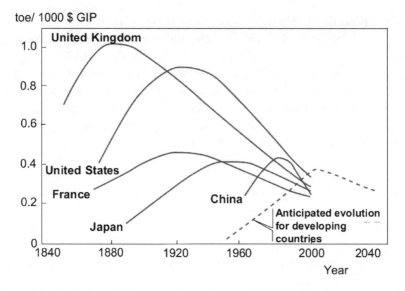

Figure 5.1   Evolution of the energy intensity (toe/1000 US $) (*Source*: Le Monde Diplomatique)

US $ GIP in 2004). However, it can be observed that during recent years this decrease has stopped [38].

Ultimately, the energy intensity of all countries tends towards a value close to 0.2 toe/1000 US $ GIP. Nevertheless, although the energy efficiency tends to decrease, the consumption of energy rises, as a result of the combined effect of demography and improving standard of life.

As an indicator, the energy intensity has to be used with care. Indeed, a reduction of the energy intensity can be presented as a success, but may be not sufficient, if it is more than compensated by the growth of the GIP. It is necessary to reach a reduction of the energy consumption in absolute terms and only as related to a permanently increasing GIP.

The value of the GIP as a development indicator is also more and more frequently challenged. It does not take into account the depletion of natural resources or environmental damage. Other indicators have been proposed. The Genuine Progress Indicator (GPI) takes into account positive and negative effects upon the environment and the society in order to correct the indications of the GIP. In the USA, while the GIP per inhabitant has roughly tripled within the last fifty years, the GPI has remained almost constant. Different trials have been made for taking into account the quality of life, by introducing indicators which take into account an index reflecting the feeling of well-being and satisfaction of the people. The Happy Planet Index is obtained by considering the product of a life satisfaction index by life expectancy and relating this product to the ecological footprint. Numerous other indicators have been introduced; the 'green GIP' was introduced by the World Bank and the Human Development Index (HDI) was introduced by the United Nations [41].

It does not seem feasible to define a single factor, which would take into account all the different factors. The existence of different indicators illustrates the fact that the GIP does not represent a unique or even the best way to measure the development of a society.

Renewing the organisation of housing and transport

The residential and tertiary sector represents 40 % of the world primary energy demand and transport 25 %.

The need to improve energy conservation requires a new organisation of housing and transport.

Dispersed suburban housing such as that which has become very wide spread in the USA implies the need to use individual transport means, which results in a higher consumption of energy and a stronger dependence on oil.

In contrast, more concentrated housing reduces transport distances and helps to develop collective transport modes. Figure 5.2 illustrates the correlation between lower energy consumption and population density [31].

Conversely, the development of individual cars leads to an uncontrolled spreading and a dislocation of the urban environment. One of the consequences is a progressive increase of the distance between the home and the work place. For example, in France, this distance was 4 km on average in 1959 and is now 15 km today. The continuity of the urban setting is also broken by the multiplication of large commercial stores, university campuses and leisure complexes.

This development is partly unavoidable, but also entails drawbacks, with consequences which are increasingly obvious. It is necessary to take into account all the different disadvantages of automotive transport (noise, pollution, time lost due to traffic jams), and the social constraints (reduced transport possibilities for elder, poorer or disabled people). Infrastructures designed to facilitate access to cars and not integrated into an overall urban development plan are unattractive and quite often destroy the harmony of a landscape, as demonstrated by numerous shopping centres in the close vicinity of large urban areas. It has been shown that an urban highway, by its detrimental impact upon local community life, can result in an increase in criminal behaviour. The city becomes a noisy and polluted area, which is left by the more affluent people who prefer to live in the suburbs.

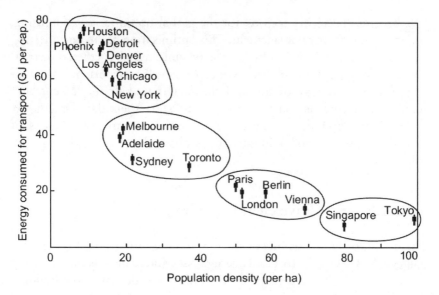

Figure 5.2   Energy consumption for transport vs. urban population density (*Source*: Newman P. and Kenworthy J., 1999)

Various attempts have been made to create more enjoyable and more sustainable urban settings. Bedzed (Beddington Zero Energy Development), constructed in 2000 south of London is the first eco-village in the world. The city of Freiburg in Bade-Wurtemberg, an area once occupied by military barracks, has been developed according to a sustainable scheme, favouring collective transport modes combined with the use of individual bicycles. In China, Dongtan, in the north of Shanghai, is designed to become the first sustainable city in the world. Its ecological footprint is planned to be reduced to a level of 2 ha per inhabitant.

In the future, the urban setting must aim at an optimal use of the available space through diversified housing, designed for limiting the time lost during transport and the consumption of energy. Rather than choosing to rely solely upon the use of individual cars, it is necessary to promote a multi-mode transport system[3], with priority given to collective transport modes for the longest distances. The development of remote data transmission and smart guidance systems should also facilitate the introduction of more flexible solutions.

---

[3] Association of different transport modes in order to use the most appropriate one for each part of the journey.

It becomes necessary to look for the global optimisation of housing, transport and energy management. This optimisation has to be undertaken at different levels: individual house or building, district ('eco-district' concept), city ('sustainable city' concept). These transformations challenge the present way of life and the fact that their implementation requires time means that it is urgent to start their deployment. As everyone is affected, it is necessary to increase public awareness of urban communities and to involve the public in the elaboration of future solutions.

Towards the positive energy building

The design of buildings must evolve by taking into account the energy balance. It is necessary to avoid energy losses and to increase solar energy inputs through the best adapted orientation and design of the building (bioclimatic architecture), and by imagining innovative solutions; thus, a vegetation cover can limit energy losses while improving the integration of the building into the surrounding environment.

New concepts of climatic housing, which have been developed, are applicable not only to individual houses, but also to large buildings. A recent design makes it possible to modulate solar energy input by using a concrete lattice, which can either let in solar light or operate as a sun visor [40]. This structure can also support solar panels.

Air circulation is achieved through natural convection and greenhouses are used to create vegetation protection of the sides most exposed to the sun's radiation (Hypergreen tower).

It is necessary to boost the efforts aimed at the development of better performing materials and equipment. This includes thermal insulation technologies, which are still advancing through the use of better performing materials and more efficient implementation. Triple-glazed windows, ensuring excellent thermal and also phonic insulation, are now available.

The reduction of energy consumption can be also achieved by using better performing equipment.

An important area for improvement is lighting, as it consumes 20 % of the world's electricity production.

Today, a fluocompact lamp produces three to four times more light per watt than an incandescent lamp. Also, the lifetime of a fluocompact lamp is around 10 000 h, that is to say, more than ten times that of an incandescent lamp. In the future, electroluminescent diodes (ELDs) will represent new options in terms of energy efficiency, with a lifetime reaching 80 000 h and a layout enabling an optimal distribution of light in buildings [43].

Energy conservation can be ensured by using better performing heating devices such as high efficiency boilers (condensing boilers), heat pumps or cogeneration.

Cogeneration and heat pumps enable a better use of electricity. In a fossil fuel power plant, electricity is produced from a heat source with a yield which, in most cases, does not exceed 35–40 %. The remaining energy is released and transmitted to the refrigeration air or water. In the case of cogeneration, the heat released is used for heating residential buildings or industrial installations.

A heat pump operates according to a different principle. It operates in a way similar to a refrigeration device. By withdrawing heat from an external medium (water or air), it supplies a quantity of heat much larger than the electrical energy consumed. The ratio of the quantity of heat thus delivered over the consumed electrical energy (coefficient of performance or COP) is frequently around 3.

Cogeneration is mainly applied in industry and in the residential or tertiary sector, while heat pumps can be used for individual houses.

In the residential area, a huge growth has occurred in terms of energy consumption. In France, in the case of an old building, dating from a year before 1975 (date of enforcement of the first thermal regulation), average thermal losses are around 330 kWh/m^2/year. The application of the recent regulation RT 2005 should lead to losses limited to 85 kWh/m^2/year.

By further improving the design of the building, it is possible to lower the level of losses down to 15–20 kWh/m^2/year, which corresponds to the 'passive' house concept. In such a case, it is possible to maintain the required temperature in the house through natural solar input and the heat produced as a result of human activities [44].

The next step (further reduction of losses and increase of external solar input) leads to the concept of the *positive energy building*, able to export energy. From the stage when heat losses do not exceed 50 kWh/m^2/year, energy provided by solar thermal or photovoltaic panels can comparatively easily exceed the heating needs of the building.

Further developments are needed to ensure a better integration in the structure of the building of thermal solar panels supplying hot water and heating and also photovoltaic panels producing electricity. The development of the equipment which is now available facilitates the integration of solar panels into the building. It becomes possible to install solar roofs or walls. Such concepts lead to the need to revise the architecture and even the structure of residential heating.

The most difficult situation occurs in the case of old buildings, which require complete restoration. For instance in France, two-thirds of dwellings were built before 1975, in the absence of any thermal regulation. The energy restoration of these buildings is presently often difficult as a result of the complexity of decision making (especially in the case of joint ownership), and the lack of the required expertise to carry out the work. It is therefore most important to deploy the incentives and regulatory measures which are needed to make the renovation of such buildings easier.

The introduction of energy quality labels and energy diagnosis represents a first step, leading to the implementation of the required transformations.

A well adapted regulation is clearly a prerequisite, even though in the long term such an investment should be profitable for the final user, due to the probable increase of the energy price.

Reduction of energy consumption in the transport sector

An initial way to limit CO_2 emissions resulting from automotive transport is to limit the use of passenger cars, by changing consumer behaviour and habits (car sharing, increased use of collective transport means, and cycling and walking instead of driving). In the case of goods transportation, besides giving a preference to local products, alternative transportation means, by rail or barge, are to be favoured.

In order to limit the consumption of engine fuels derived from oil, it is also necessary to reduce the consumption of vehicles per unit distance covered. Such a reduction of the consumption can be achieved through two complementary pathways:

- *Reduction of the energy required for moving a vehicle.* The quantity of energy needed to move a car can be reduced by limiting its weight and by friction losses [45]. Unfortunately, recent developments have gone in the opposite direction in order to meet the various expectations of consumers (increased comfort and security).
- *Improving the efficiency of the power train.* The main problem arises from the difficulty in combining a further increase of the efficiency with more and more stringent regulations concerning pollutant emissions.

Different technologies are investigated. In the case of a spark ignition engine, these include mainly 'downsizing' with turbo-charge of the engine, stratified combustion combined with gasoline direct injection and homogeneous combustion. The basic principle of this new combustion mode consists of the preparation of a much more homogeneous air–gasoline mixture in the combustion chamber. By getting a more homogeneous temperature distribution and avoiding local temperature peaks, it is possible to reduce sharply the formation of pollutants, namely nitrogen oxides and soot. Homogeneous combustion can be used also in the case of a diesel engine making it possible to maximise the yield, while reducing the emissions of pollutants such as particles and nitrogen oxides (NO_x) [46].

European passenger car manufacturers initially committed themselves to reduce the emissions of CO_2 per covered kilometre, from 190 g/km in 1997 to 160 g/km in 2003, 140 g/km in 2008, and with the objective of 120 g/km in 2012.

Figure 5.3 illustrates the evolution of passenger car CO_2 emissions within the European Union (UE 15). For about ten years, these emissions have decreased at the rate of 1.4 % per year. To achieve the objective for 2012, it would be necessary to have a decrease in rate of 3.3 % per year in the years to come.

It is apparent that these objectives will not be reached and the level of constraints has been reduced. They are especially difficult to reach in the case of spark ignition engines, due to efficiencies which are on average lower than in the case of diesel engines. The objective for 2012 is now fixed at 130 g/km, a further reduction of 10 g/km being expected as a result of other means (reduction in the carbon content of engine fuels). The weight of the vehicle is also taken into account.

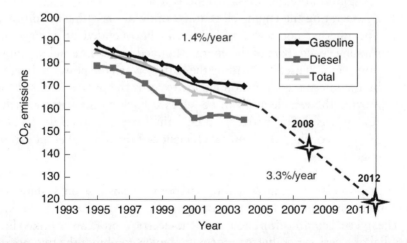

Figure 5.3 Evolution of passenger car CO_2 emissions within the European Union (*Source*: IFP). Reproduced with permission from *Energie & Climat: Réussir la transition énergétique* by Alexandre Rojey, Éditions Technip, Paris, 2008

The difficulties encountered when trying to reach these objectives have pushed car manufacturers to investigate the use of alternative fuels emitting less carbon of fossil origin per unit of energy produced (biofuels, NGV).

Hybrid propulsion

One further option for reducing passenger car CO_2 emissions is the use of hybrid propulsion. The hybrid propulsion concept for passenger cars was initially introduced in 1997 by Toyota in the Prius car.

It consists of using the association between a thermal engine and an electric engine for ensuring the propulsion of the vehicle. A battery for electricity storage is required to supply the energy needed by the electrical engine.

Using a hybrid propulsion system makes it possible to reduce energy consumption:

1. By making it possible to stop the thermal engine, when the vehicle is stationary rather than when the engine is running ('stop and start' function). This function requires only a limited power increase of the starter as compared with a standard one and yet makes it possible to achieve a reduction in the energy consumption of 7–10 %.
2. By recovering part of the energy lost when braking; the additional consumption reduction is around 4–8 %.
3. By operating the engine close to its optimal operating conditions. The internal combustion engine can be operated at a constant running speed. Part of the energy supplied by the internal combustion engine is used for loading the battery when the power delivered by the internal combustion engine exceeds the power required for moving the vehicle and can be released back by the battery in the reverse case. The corresponding gain varies between 10 % and 30 %, according to the initial efficiency of the internal combustion engine.

Hybrid propulsion requires a more elaborate transmission system for making the most of all these new options.

The internal combustion engine and the electric motor can be linked in a parallel or a series–parallel arrangement, which combines the two possible driving modes. Further possible options arise from the possibility of operating the system in a purely electric mode by using plug-in batteries, which can be recharged by using an external electricity supply source. Such an option will be discussed in Chapter 6.

Hybrid propulsion is typically a transition technology. It helps to significantly improve the performance of the present vehicles, while leading the way for new propulsion systems [42]. It contributes to the acceleration of further progress in the area of electric batteries and should thus facilitate the development of electric cars.

Energy storage

Energy storage represents a key technology for adjusting energy supply and demand, the energy being stored when the availability is higher than the need and released back in the reverse situation.

It makes it possible to adjust the energy supply to a variable demand from a source delivering a constant power. Conversely, energy storage makes it possible to adjust the energy supply to a constant demand from an intermittent delivery source.

Fossil fuels have the advantage of storing large amounts of energy within a comparatively small volume. For renewable energy sources this is not the case, and a large deployment of these renewable energy sources requires the use of storage systems to compensate for the intermittence of the supply and also for mobile applications (on board storage).

In the case of electricity production, flexibility can be ensured through the modulation of the power delivered by a fossil fuel power plant operating as a back-up.

Thus, a natural gas fired combined cycle can be used to compensate for the intermittence in electricity production supplied by windmills. Such an option limits the share of renewable energy sources which can be introduced in the electricity generation system.

Hydraulic storage is the only method currently used on a large scale. In France for instance, the hydraulic storage of Grand'Maison in the Alps has a storage capacity of around 400 TWh, handling a peak power amounting to 1.2 GW when storing and 1.8 GW during discharge. Unfortunately, most natural storage sites of this kind are already exploited, at least in Europe, and it is not easy to extend the storage capacity of such hydraulic systems. An option to be explored might be the development of new sites by creating artificial lakes.

Figure 5.4 presents a comparison between different types of storage systems and illustrates the difficulty in finding a good alternative to liquid hydrocarbons [47]. The specific energy stored in liquid hydrocarbons

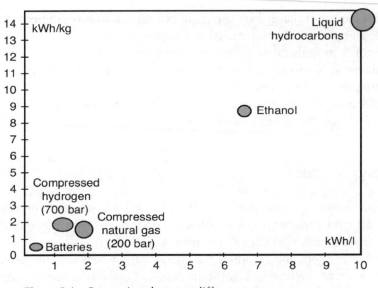

Figure 5.4   Comparison between different energy storage systems

amounts to 12 kWh/kg, whereas in the case of storage of hydrogen at a pressure of 700 bar, the stored energy density does not exceed 1.2 kWh/kg.

The electrochemical energy storage density provided by a battery is even lower by an order of magnitude (20–120 Wh/kg). In the case of mechanical systems such as flywheels or compressed air storage, the stored energy density is also comparatively small: 1–5 Wh/kg for flywheels and 8 Wh/kg for air compressed at 200 bar. Still, electricity or mechanical energy storage should not be considered as directly equivalent to the chemical energy storage provided by a hydrocarbon. A hydrocarbon delivers energy in the form of heat, which has to be converted into mechanical and then electrical energy with an efficiency substantially less than 1. Furthermore, the energy is generated in an irreversible way and the system cannot be reloaded.

Although the specific energy density is comparatively limited in most cases and especially when mobile applications are considered, electrochemical battery storage is the most readily applicable. Extensive development programmes are underway to further improve the different types of batteries which can be used (lead-acid, nickel and lithium). Nickel-metal hydride (NiMH) batteries are most commonly used in the presently available hybrid cars. The specific energy storage density of these batteries (55 Wh/kg) is significantly higher than the storage density of lead batteries (25 Wh/kg). Lithium batteries (lithium-ion and lithium polymer) are considered the most promising (70 Wh/kg), but are comparatively expensive. Sodium-sulphur batteries are also being investigated for large capacity storage applications. These batteries operate at a high temperature (350 °C). They use liquid sulphur and sodium electrodes, separated by a solid electrolyte. The storage capacity of such batteries is around 100 Wh/kg [72].

Liquid phase electrochemical storage systems, developed recently, look promising. They include a reversible battery, equipped with two compartments separated by an ion exchange membrane.

Each compartment is connected to an electrolyte reserve. Each electrolyte is pumped through a circulation circuit and, while one of the electrolytes is oxidised at the contact of the membrane, the other one is reduced [73]. The system is reversible and is thus able to store energy. One of the most attractive electrolyte couples is based on the use of vanadium ions and uses the couples V^{2+}/V^{3+} and V^{4+}/V^{5+}. Three devices operating according to this concept, with a storage capacity ranging between 500 kWh and 2000 kWh, have been installed.

Electrochemical storage systems cannot be used for storing the amounts of energy needed to ensure a seasonal adjustment between supply and demand. Presently, only gravity hydraulic storage can be used on such a scale.

Underground storage (saline cavity or aquifer) represents another potential way of achieving a large storage capacity (compressed gas, heat, hydrogen) and might be more widely used in the future. Underground compressed air storage has already been implemented. It is thus possible to store 12 kWh/m^3 of air at 100 bar. At Huntdorf, in Germany, air compressed at 70 bar is stored in two caverns representing a volume of 310 000 m^3. In the future, large quantities of hydrogen might be stored in underground saline aquifer systems, similar to those which are used for storing natural gas. It is also feasible to store hot water underground at comparatively shallow depths.

In the residential area, the use of thermal storage materials helps to recover heat during the hottest hours of the day and to release this extra heat during the coldest hours. Different materials can be used for such a purpose and in this area too, innovative technologies are being investigated, including the use of phase transition materials. Such materials can store excess heat during the day and release this heat during the night. These heat storage systems are especially useful for buildings supported by a light infrastructure.

Energy storage is certainly a key technology for the energy transition period, and it is essential to make it more cost effective during the years to come.

New agricultural production modes

To provide the food needed by the world's 9 billion inhabitants expected by 2050, it would be necessary to double the present agricultural production.

The green agricultural revolution, which began after the Second World War, has led to a strong increase in agricultural production, thus helping to avoid famines in numerous regions worldwide and especially in Asia. However, this revolution has been accomplished by using high amounts of energy, fertilisers and pesticides. Similar to what has occurred in other economic sectors, this has had a negative impact upon the environment and has increased the dependence of agricultural production upon energy supplies.

Therefore, new agricultural production modes have to be developed. A first pathway consists of applying a more intense selection of plant varieties, genetic engineering methods and new chemical products in order to increase productivity further.

In order to reduce the consumption of energy and the impact upon the environment, alternative more ecology-friendly methods are also being investigated. These methods aim at the optimisation of the ecosystem and avoid ploughing up the ground. The objective is to increase productivity, while ensuring better energy conservation, by using new cultivation methods involving reduced external inputs, combined with new rotation and irrigation methods [48].

The use of genetically modified organisms (GMOs) should be examined in the most objective way. It would be unfortunate to discard without a thorough investigation new options which might satisfy the food needs of the world's population in a better way, by consuming less energy or by using the available resources (especially water) better.

In this area also, rather than looking for a unique solution, which would provide answers to a whole set of raised questions, it seems preferable to favour a pragmatic approach, taking into account the real needs of the population and environment protection.

Sustainable agriculture can help also to supply a more healthy food. In developing countries, it generates new jobs. It is essential to protect such sustainable agriculture in competition with products derived from energy intensive agriculture by taking into account such factors within the international trading mechanisms.

Consumption habits are also very important. The average cereal consumption per inhabitant in the USA is four times higher than in Japan (800 kg/year against 200 kg/year). This is mainly due to a much higher consumption of meat. However, life expectancy in Japan exceeds that in the USA by 8 years [7]. This does not mean that consumption habits in Japan are better in all respects as they contribute to the depletion of world seafood resources, but this example shows that health and life expectancy criteria go often hand in hand with reduced energy consumption.

It is most important to succeed with a transition in the agricultural sector as it affects some of our essential needs, namely food and health. Progress should lead to a reduction in the consumption of energy and of all our natural resources, while reducing the pollution resulting from the use of chemicals.

More efficient industrial processes

Industry optimises continuously the processes it uses in order to adapt to the external economic and regulatory signals it receives. In the years to come, it will have to adapt to increasing energy prices and also to a reduction of CO_2 emissions resulting from the allocation of quotas and the development of the emissions trading scheme.

In order to fulfil that goal, industry will use more efficient processes. A wide range of options is available:

- Better energy integration and limitation of energy losses (e.g. introduction of better thermal insulation).
- New generation, more efficient processes (e.g. better performing catalysts, membranes or more selective adsorbents).
- Introduction of energy recovery devices (heat exchangers, expanders).
- Online control systems for optimising the use of energy in a continuous way.

The development of future industrial processes will benefit from all the progress accomplished in the area of materials (composite materials, nanotechnologies, etc.), information technologies and biochemistry. The convergence of progress from these different areas will result in the development of new manufacturing processes which perform better and are less energy intensive.

Waste treatment and materials recycling

Processes and equipment ensuring better energy conservation should be considered as contributing to the development of green technologies, which aim at protecting the environment and preserving natural resources (energy, matter, raw materials).

Green technologies are to be found in all sectors of the economy. They also include antipollution technologies that are used to tackle any air, water and soil pollution.

An important application area for green technologies is waste treatment, in order to avoid polluting the environment and to recycle the maximum amount of raw materials which are used.

Wastes, which represent a risk for the environment, can therefore become a resource:

- They can be used for generating energy, either directly in incineration plants, or after transformation into a fuel (biogas, pyrolysis oil or second generation biofuel).
- Raw materials can be recycled; while avoiding the pollution which would result from dumping the waste in the environment, this recycling helps to preserve the raw material thus recovered. It also represents a way to reduce energy consumption, as the amount of energy required for recycling the material is usually lower than the amount required for the initial transformation.

Recycling of raw materials is already widely practised. The supply of iron for the production of cast iron and steel was ensured in 2004 by using 1100 Mt iron ore supplying 695 Mt of iron and 400 Mt of scrap iron supplying 300 Mt of iron. Excluding CO_2 emissions resulting from the generation of the electricity needed for the transformation process (which depends upon the energy source), the quantity of CO_2 generated is 2200 kg per ton of steel in the case of a production pathway starting from iron ore (blast furnace and oxygen converter) and only 100 kg when starting from scrap iron. This means that emissions reduction is huge especially if using electricity with low carbon content [49].

A better use of resources is to be based both upon recycling (or 'circular economy') and 'functionality'. Functionality consists of supplying a service rather than selling equipment, which results in better use of the equipment and necessary resources while recycling any used component [50].

Research actions required

... **For reducing energy consumption while protecting the environment**

- **New better performing and lower energy content materials**
  - Insulation and phase transition materials
  - Highly resistant composite materials, for reducing the weight of vehicles (road or air transportation)
- **Thermal systems**
  - Heat exchange systems
  - Thermodynamic cycles for upgrading low temperature heat
- **High efficiency internal combustion engines**
  - New combustion processes
  - Hybrid power train
- **Energy storage**
  - High performance batteries
  - Underground energy storage (compressed gas, heat, hydrogen)
  - Electrochemical liquid phase storage
- **Green technologies**
  - Analysis and characterisation of pollutants (in the atmosphere, water and soil)
  - Treatment of industrial emissions
  - Treatment of polluted air, water or soil
- **Waste treatment and recycling**
  - Energy production
  - Recovery and recycling of raw materials

6

Reducing the Energy Carbon Content

The carbon content of the primary energy supply

The primary energy supply comprises all the energy sources which contribute to the overall supply of primary energy: nuclear energy, coal, oil, natural gas, hydroelectricity, biomass, wind, solar and geothermal energy.

The diversification of energy sources is essential. It helps to reduce the dependence on oil imports and thus to promote the security of supplies. However, it is not necessarily beneficial in terms of climate change. Thus, in the absence of appropriate measures, the comeback of coal might lead to a significant increase in CO_2 emissions.

From this point of view, the situation can be improved only by increasing the share of low carbon energy (nuclear and renewable), or by using as energy vectors electricity or hydrogen produced from fossil fuels with CO_2 capture and storage (see Chapter 8).

It is also possible to use biomass as an energy source. As the carbon contained in biomass has been extracted from the atmosphere during the photosynthesis process, it is considered as recycled when it is emitted as a result of biomass combustion. The carbon balance of biomass combustion is therefore carbon neutral (except if fossil energy has been used for biomass production). A complete carbon balance has to take into account all CO_2 emissions resulting from biomass production (agricultural machines, fertilisers, etc.), harvesting and transport to the final user site.

Energy and Climate: How to achieve a successful energy transition   Alexandre Rojey
Copyright © 2009 Society of Chemical Industry

When energy is produced from a fossil fuel, the quantity of CO_2 emitted per unit of energy produced depends on the hydrogen/carbon ratio of the initial fuel. The energy carbon content can be expressed in kg of CO_2 (or only carbon) per unit of energy produced (GJ). The carbon content of natural gas is therefore much lower than that of coal, as shown in Table 6.1.

Table 6.1 Comparison of CO_2 emissions for various fossil energies (*Source*: ADEME)

	Heating value (GJ/t)	CO_2 emissions per unit mass (t/t)	CO_2 emissions per toe (t/toe)
Natural gas	57 (LHV)[a]	3.26	2.394
Oil	42	3.066	3.066
Coal	26	3.14	3.990

[a] The lower heating value (LHV) or net heating value does not take into account the heat supplied through the condensation of steam contained in the flue gases.

In addition, a fuel such as natural gas offers high energy efficiencies, especially through the use of combined cycles. In this type of cycle, steam is produced by recovering heat from the exhaust gases of the gas turbine and used to drive a steam turbine. The energy efficiency of a natural gas combined cycle therefore reaches a value of nearly 60%, whereas the efficiency of the present coal-fired power stations is in the range of 45–47%.

By substituting natural gas for coal, the CO_2 emissions per unit of energy produced can therefore be reduced by a factor of 2–3. Substitution of biomass for coal can lead to even better results, since the carbon emitted is then considered as recycled (provided, however, that the CO_2 emissions associated with the production and transport of biomass are sufficiently reduced).

Carbon intensity

Carbon dioxide emissions per inhabitant in the world are highly variable. They range from 1 tonne per year per inhabitant for India up to 18 tonnes per year per inhabitant for North America.

They are directly correlated to the Gross Domestic Product (GDP) per inhabitant, as shown by the graph in Figure 6.1. Some countries are above the correlation interval (United States, Canada, Russia, Saudi Arabia, Poland, South Africa). In contrast, France and Japan, which are below the interval, may be considered as comparatively 'virtuous'.

The carbon intensity is defined as the level of CO_2 emissions related to the GDP. Between 1990 and 2005, the carbon intensity dropped in several of the large economic regions, changing in Europe from 0.5 to 0.4 kg of CO_2 per US \$ and in North America from 0.7 to 0.5 kg of CO_2 per US \$.

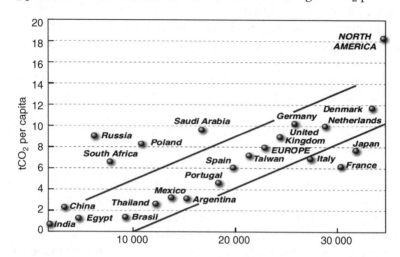

Figure 6.1   Relationship between CO_2 emissions and GDP per capita in the world (*Source:* Enerdata)

In China the figure dropped considerably, from 1.3 to 0.7 kg of CO_2 per US \$. Nevertheless, it remains substantially higher than that of the main industrialised countries. In India and Latin America the figure remains stable at about 0.4 kg of CO_2 per US \$. In Asia in particular, it can be observed that the carbon intensity starts increasing when the GDP per inhabitant rises, then reaches a maximum before decreasing.

The same care must be taken with carbon intensity as with energy intensity. A drop in carbon intensity cannot be considered as a sufficient result per se. An actual drop in emissions is what is really needed.

Energy carbon content reduction pathways

Energy decarbonation can be achieved either by using an energy source of non-fossil origin or by capturing and storing the CO_2 emitted, to tend towards a neutral carbon balance.

The first option, which is the most direct, consists of using either nuclear energy or the renewable energies.

These two alternatives to the fossil energies will be examined in the remainder of this chapter. In most cases, they produce electrical energy, which results in certain restrictions in their use, since electricity does not cover all requirements and is difficult to store.

In numerous applications, these energies cannot be directly substituted for the fossil energies, whether to replace the fuels used in transport or to produce energy upon request.

One solution to meeting these requirements consists of using the energy produced from biomass, as energy production from biomass can be considered as carbon neutral.

Biomass stores energy in chemical form, like the fossil fuels, and can be used in a boiler or to produce a biofuel used in an internal combustion engine.

Capturing and storing CO_2 is also a way to achieve a zero carbon balance. This option will be analysed in more detail in Chapter 8. In the specific case of biomass, by capturing and storing CO_2 emitted during combustion, it is even possible to achieve a net gain in the carbon balance.

It is thus possible to produce hydrogen or biofuels with negative carbon content. Such an option is not currently used but might become attractive in the future for improving the carbon balance.

Hydrogen generates energy by combustion without emitting CO_2. However, it requires an energy source to be produced. It is only an energy vector and its interest in terms of carbon balance has to be evaluated by considering all the stages from its initial production to its final use.

Development of new energy vectors, such as hydrogen, is a way of promoting the use of alternative energies with zero or negative carbon content, by extending the range of applications in which they can be used.

Lastly, widespread penetration of these alternative energies cannot take place without a significant increase in the number of energy storage facilities, energy currently being stored by the fossil fuels.

Revival of nuclear energy

The share of nuclear energy in the worldwide energy balance is currently limited due to its high technical requirements and the cost of the investments needed. In many countries, its development has been hindered by generally adverse public opinion, in view of the inherent hazards. The rise in the price of oil (and correspondingly of natural gas) creates a climate which is much more favourable to the revival of nuclear energy.

Currently, all nuclear power stations use the energy released by fission of heavy isotopes such as uranium 235 (^{235}U)[1]. After the first nuclear

---

[1] Uranium 235 is a fissile isotope of uranium. Natural uranium deposits contain 0.7% $U235$.

power stations (first generation), the current series of pressurised water reactors (PWRs; second generation), such as those implemented in France, have been successfully operated for nearly 50 years. With PWRs, the heat produced in the nuclear reactor core is sent to a primary coolant loop and then used to generate steam by heating a secondary water loop in a steam generator. With boiling water reactors, steam is generated directly in the reactor core. Although this avoids the need for an additional loop, the core volume is larger. Most of the 440 nuclear reactors currently operating in the world (including 58 PWRs in France) are one of these two types. A nuclear power station consists of one or more 'generating units', with each generating unit having a reactor. For the large current power stations, the power produced by a generating unit varies between 1300 MW (1.3 GW) for the present reactors and 1600 MW (1.6 GW) for the future European Pressurised Reactor (EPR).

A PWR type power station operates with enriched uranium containing about 4% ^{235}U. The uranium is enriched by gaseous diffusion or, more and more frequently, by ultracentrifugation.

Development of a new generation (third generation) of PWRs with improved safety should facilitate the revival of nuclear energy. The commissioning of a reactor of this type in Finland, and another at Flamanville in France, will represent the first industrial references of this process.

New power station projects are planned in the USA and China. Great Britain has also decided to reconsider this option.

This revival will involve the development of new processes, which make better use of uranium resources. At the current rate, the world uranium resources identified to date (4.75 Mt of natural uranium) can supply the 450 reactors currently in service for more than 70 years. If nuclear electricity production develops rapidly, it will create significant tension over uranium supplies. The price of uranium is already rising sharply: in current monetary terms, from the start of 2001 to the end of 2006, it has escalated by 1000%.

If production increases rapidly, this tension over uranium is likely to increase. In particular, the uranium resources identified would not be sufficient to cover the demand from 2020 under the *World Nuclear Association* 'high' scenario and resources not yet discovered would be required [54].

The current electricity-generating reactors which use energy from the fission induced by thermal neutrons and consume mainly uranium 235, harness only about 0.6% of the energy potentially contained in the natural uranium.

In the longer term, by 2030–2040, fast neutron reactors (fourth generation) will be developed [52, 53]. Fast neutron reactors operating as 'breeder reactors' can use uranium 238, producing fissile plutonium (isotopes 239 and 241) by neutron capture on uranium 238. By successively recycling the plutonium produced from the uranium 238, 70–90% of the initial uranium can be fissioned, depending on whether or not the minor actinides also produced are recycled. Uranium resources could then be used for a period at least fifty times longer compared with the current situation [54–56]. The investment required for a fast neutron reactor (FNR) is higher than for a PWR, but the cost of the kWh produced becomes almost independent of the price of natural uranium. A certain number of technical problems still remain to be examined and the reliability of the process must be demonstrated before large-scale industrial deployment. One problem is the resistance of the materials, in particular steel, under high irradiation.

We can expect to see third generation reactors widely deployed as of 2010, while a large proportion of the nuclear reactors in service throughout the world will remain second generation for at least the next two or three decades (these reactors are designed for a lifetime of 40 years, possibly extended to 50 years on a case-by-case basis). Under these conditions, commissioning of fast reactors (fourth generation) could start around 2040–2050, but at this time numerous third generation reactors will still be in service [56, 57].

Thermoelectric fusion is a pathway which is also being explored, with the construction of the ITER experimental facility in the south of France, but prospects of success are uncertain.

Current projects implement deuterium-tritium fusion. Tritium can be produced from lithium 6 available in smaller quantities than deuterium, but this type of process could nevertheless produce energy for several thousand years. Deuterium-deuterium fusion could also be implemented, but the problems to be tackled are even more difficult and represent a consideration for the future. Industrial applications of thermoelectric fusion can only be expected around the end of the century or during the next century. The stake remains considerable, however, since it offers the possibility of accessing a form of energy so far untapped and of immense potential.

During the energy transition period, we expect that only the fission processes will have a real impact. These processes offer substantial advantages due to absence of CO_2 emissions during electricity production and the financial profitability, which is becoming increasingly attractive as the price of fossil energies increases.

Nuclear development is held back first by the investments required and secondly by the reluctance of public opinion in various countries, faced with the actual or assumed dangers inherent to the production of nuclear energy.

The investments required are initially higher by a factor of 1.5 to 2 compared with a coal-fired power station and by a factor of about 3 compared with a gas-fired power station. They represent an obstacle whose impact on the final cost of the kWh produced increases as the production cost taken into account increases. The rate of use at full power is also an important factor, as well as the total lifetime, which may reach 40–60 years.

The main limitations are nevertheless due to the fears still generated by nuclear power. The risks put forward concern the safety of the power station itself, the risks of proliferation related to management of the fuel cycle[2] and those related to the storage of radioactive waste. The fears, which originally mainly concerned the safety of the power station itself as well as the effect of long-lived radioactive waste, are now shifting to the risks of nuclear proliferation, especially during the producing of enriched uranium in centrifuges, as illustrated by the situations in Iran and North Korea.

Even through suitable solutions may be provided, especially through international inspection agreements, the fact remains that it will take a long time to substantially increase the proportion of nuclear energy.

The problem of waste is also a delaying factor. Low-activity waste (waste 'A') must be stored and monitored during periods extending from two to three centuries. The most active waste (waste 'C') must be confined for very long periods of time. The option chosen in France consists of confining the waste in so-called 'reversible' deep geological storage: it can be retrieved after one or two centuries.

The plutonium contained in the fuel used is sometimes considered as waste and sometimes as a resource. In France, it is separated for recycling as 'Mox' fuel. Since this operation cannot be repeated, however, stocks of plutonium are increasing throughout the world. It is planned to reuse this spent fuel in 'fast breeder reactors' but, as we have seen, their large-scale deployment is not expected for many years. This means that large quantities of spent fuel containing plutonium must be stored.

---

[2] This terminology, commonly used, can be misleading as energy is produced by nuclear fission and not by combustion.

In its reference scenario [1], the IEA predicts a drop in the share of nuclear energy in electricity production from 16% in 2006 to 13% in 2030, despite an increase in installed power from 372 GWe in 2007 to 433 GWe in 2030.

In view of the advantages of nuclear energy in terms of climate change, however, we could witness much faster growth of nuclear energy in the future. According to a vision from the nuclear industry, the 'high' scenario published by the *World Nuclear Association* in 2005 predicts an installed power of 740 GWe, which amounts to a reversal of the current trend.

This nevertheless shows that, even assuming a sharp increase in the share of nuclear energy in electricity production, it will only be possible to cover a limited fraction of the primary energy supply throughout the transition period.

Renewable energy prospects

Renewable energies are currently considered as a mere backup solution. In the world, they represented 1513 Mtoe in 2006, i.e. 13% of the primary energy production [1].

Out of this total, biomass and waste accounted for 1186 Mtoe and hydraulic power 261 Mtoe. These two energy sources, which make the largest contribution to the renewable energy balance, have already been used for many years.

The potential for development of hydraulic power remains limited. For the last thirty years or so, the contribution of hydraulic power in worldwide energy production has stagnated in absolute value and its relative share has dropped, from 21% in 1973 to 16% in 2006. In France, construction of new sites mainly concerns mini or micro hydraulic facilities, but it is faced with numerous obstacles, not only financial but also in terms of social acceptability.

The case of biomass energy is quite different. Biomass represents a considerable potential, with a worldwide estimated figure of 2230 Mtoe, including 1600 Mtoe of forest origin [58].

The biomass energy development prospects are therefore promising, especially as regards cogeneration of electricity and heat. They will be analysed below. The case of biofuels, another important market for biomass, will also be examined in further detail below.

The main development hopes concern wind and solar power, which still only represent a minor proportion of energy production (respectively, 0.34 and 0.09 Mtoe in France in 2007). The disadvantage with these energies, however, is that they are intermittent and diluted. On account of being intermittent, their contribution can only be limited unless an expensive storage system is installed.

Wind energy: an already mature energy

Wind power technology is already relatively mature, producing electricity under almost profitable conditions. Current machines can develop powers from 1.25 to 2.5 MW. The rotors fitted on 2.5 MW machines have a span of up to 80 m [59].

Virtually all high-power wind turbines have a three-blade rotor, which offers higher efficiency than two-blade rotors, without making construction of the wind turbine overcomplicated.

Wind energy is almost competitive, the cost price of the electricity produced being in the region of € 50/MWh. The main drawback with wind energy lies in the fact that it is intermittent and cannot be modulated to match demand. Intermittent operation must be compensated by the grid, which limits the share of electricity that can be supplied by wind energy.

In the future, the price of fossil fuels, together with constraints on CO_2 emissions, should make wind energy more competitive.

Recently, the development of wind energy, especially within the European Union, has been supported by a voluntary policy, imposing an attractive buy-back price. The installed power has therefore increased considerably over the last few years, especially within the European Union.

The global figure reached 74 GW in 2006 and 94 MW in 2007. The installed power in Germany reached 22 GW in 2006. The two other European leaders are Spain with 11.6 GW and Denmark with 3.1 GW installed in 2006. In the USA, the installed capacity in 2006 was at the same level as in Spain.

We can expect to see offshore projects developing in the future, despite higher costs, to take advantage of stronger and more regular winds and also to avoid the visual and sound nuisance associated with land-based constructions. Denmark already has an offshore installed power of 400 MW. Germany is considering the installation of offshore sites which will produce 10 GW in 2015 and 20 GW in 2020.

In addition to wind energy production at sea, there is a revival of interest in the possibility of using sea movements (currents, waves), a field which has been studied for many years. There are various ways of harnessing the energy from the sea:

- Mechanical energy of the waves, tides or sea currents exploited by water turbines similar to wind turbines, but located under the sea.
- Thermal energy of the seas in the inter-tropical regions, based on the temperature difference between surface water and deep water.
- Energy of the salinity gradients based on the difference in salinity between fresh water and sea water.

Due to the difficulties in producing and transporting the energy produced, developments in this field have not gone beyond the stage of design, isolated facilities (the Rance tidal power plant) or prototypes (wave energy and water turbines). For the time being, offshore production of wind energy seems to be the only option feasible on a large-scale basis.

The promise of solar energy

Solar energy is the main renewable energy resource throughout the world. Other renewable energy sources, e.g. biomass energy and wind energy, are derived directly from it. It is an abundant energy source. Our planet receives from the sun the equivalent of 15 000 times the energy consumed in the world, but this energy is diffuse and intermittent.

The power received at noon with no cloud cover is about $1\,kW/m^2$. Given the day–night intermittence and weather fluctuations, the energy received in 24 h is between $2\,kWh/m^2$ and $3\,kWh/m^2$ in northern Europe, and between $4\,kWh/m^2$ and $6\,kWh/m^2$ in southern Europe or between the tropics.

Solar energy can be captured as either heat or electricity using the photovoltaic effect.

There is considerable scope for the development of low-temperature thermal solar energy in the short term. Heat is supplied by solar sensors consisting of a black absorbent surface which transfers the heat to a heat exchange fluid, generally a mixture of water and glycol to prevent the possibility of freezing. A glazed surface is fitted over the absorbent surface to block the infrared radiation re-emitted.

Selective coatings such as chromium oxide deposited on copper are used to reach temperatures of 70–$90\,°C$, in order to limit the re-emission of infrared radiation. To reach temperatures of more than $100\,°C$, sensors under vacuum are necessary, in particular to supply absorption solar air conditioning systems [64, 65].

In the housing sector, thermal solar energy is used mainly to provide sanitary hot water. It may also be used to cater for a certain proportion of heating requirements. For these applications, flat sensors have efficiencies in the region of 50%. An area of about $4\,m^2$ of sensors is required to meet the hot water requirements of a family of four [64].

High-temperature thermal solar energy to produce electricity, requiring concentration sensors, is not yet profitable and its future prospects are still a subject of debate.

Photovoltaic electricity generates high hopes, although it is not yet directly competitive with the electricity produced in the current power stations. Consequently, the only present applications concern isolated sites for the supply of relatively reduced levels of power. Significant progress has nevertheless been observed. As with wind power, incentive policies can be set up to offset the cost of developments on a larger scale. In particular, the possibility of exporting some of the electricity produced to the electricity grid, at a sufficiently attractive purchase price, should promote deployment of photovoltaic panels in the housing sector. As mentioned in Chapter 5, integration of solar panels in the housing sector could revolutionise the future design of buildings.

Photovoltaic cells are assembled in modules. After a significant drop, the average cost of the production modules is currently tending to stagnate or even increase slightly due to a shortage of silicon, further to a rapid increase in demand. The cost of a system connected to the grid is about € 5/Wp[3]. For a system equipped with battery storage, this cost is between € 6/Wp and € 8/Wp. The cost of the kWh produced is between € 0.25 and € 0.5 without storage and between € 1 and € 1.5 with storage [61].

The global installed power increased from 20 MWp in 1985 to 37 500 MWp in 2005 [60]. Current forecasts predict an installed power of 66 400 MWp in 2020.

The market is currently dominated (80% share) by mono- or multi-crystalline silicon cells. We can expect further progress in the field of photovoltaic cells, especially by reducing the thickness of the silicon layers and by mass production. Thin layers are produced using multicrystalline silicon deposited on various substrates (10–$40\,\mu$m layers) or amorphous silicon ($1\,\mu$m layers).

---

[3] Wp, watt of peak power. The peak power is the maximum power which can be delivered when solar radiation is at a maximum.

However, while the conversion efficiency of monocrystalline silicon cells is approximately 16–17%, the figure drops below 10% with multi-crystalline silicon thin layers. The efficiency of amorphous silicon cells is even less, which accounts for the fact that their market share has become very low. Mixed structures combining crystalline and amorphous silicon seem to offer a potentially interesting option.

Studies are being conducted on other materials such as copper-indium diselenide, from the family of chalcopyrites, which offers the possibility of high efficiencies, nearly 20%. Semiconducting organic materials may also be used. They are easy to implement; they can be used in the form of flexible sheets but, for the time being, their efficiency is at best about 5% and their lifetime is insufficient [62–64].

As prices continue to drop, photovoltaic electricity production should eventually become competitive.

Before we can expect to see widespread development in the use of photovoltaic cells, progress is required in the field of energy storage to overcome the problem of intermittent energy supply.

Advantages and limitations of geothermal energy

The principle of geothermal energy is based on extraction of heat from the subsoil.

Low-temperature geothermal energy is easier to obtain: it is used as an additional energy source when heating buildings, generally being associated with heat pumps. The energy supplied for heating purposes reached 3 Mtoe in 2006.

Geothermal energy can also be used for the production of electrical energy in countries with a high geothermal gradient, such as Iceland. Provided that the drilling costs can be reduced sufficiently and the necessary materials are available, this type of energy production could be extended to other countries by implementing deep geothermy.

In 2006, electricity production represented the supply of 59 TWh, for an installed power of about 9 GW.

Compared with solar and wind, geothermal energy has the great advantage of being supplied on a continuous basis. Expertise and knowledge of the subsoil are required, however, hindering its development. The necessary infrastructures must be paid back over relatively long durations, with the risk of a temperature rise of the geothermal source after a long period of operation, since the heat removed is only renewed very slowly.

Biomass energy and biofuels: potentialities and risks

Biomass is a renewable energy source, which presents the advantage of storing energy.

We can expect to see an increase in the role of biomass for energy production in the future. Biomass offers the advantage of being a renewable energy which is also storable. In the European Union, energy from biomass reached 61 Mtoe in 2006, an increase of 3.1 Mtoe compared with 2004. In the world, biomass represents the main source of renewable energy, amounting to 1186 Mtoe in 2006 [1, 66].

The production of heat by combustion represents the main energy use of biomass, wood being the primary renewable energy source in Europe. Wood can be burnt in individual or collective boilers. It can also be used to produce steam, generate electricity as well as heat in cogeneration.

Anaerobic methane fermentation in the presence of waste rich in organic matter produces biogas. At least some of the methane produced can also be recovered in open refuse tips by *aerobic* fermentation. The main disadvantage of the biogas produced in this way is that it contains numerous contaminants, and in particular highly corrosive acid compounds. In general therefore, biogas requires intensive treatment. Biogas can also be used as a compressed gas fuel (NGV). Although this application is faced with the difficulty of distributing a gaseous fuel, it offers a highly interesting environmental balance.

Production of biofuels currently represents the main alternative to petroleum fuels in the field of transport. The production of biofuels has grown very rapidly, generating strong controversies about its negative impact on food supply. The world production in 2006 reached 24.4 Mtoe, as compared with 10.3 Mtoe in 2000.

Biofuels offer the advantage of reducing the dependence of the consumer countries on oil while at the same time improving the CO_2 balance. The CO_2 emitted by combustion of biomass is seen as neutral with respect to the greenhouse gas balance since it can be considered as being recycled during photosynthesis, as indicated earlier. We must nevertheless take into account all emissions generated during the production, transport and transformation of biomass (life cycle analysis), which may in some cases significantly reduce, or even completely cancel out, this advantage.

The European Union member states have set an initial goal of incorporating at least 5.75% of biofuels in 2010 and 10% by 2020 in the fuels of fossil origin (gasoline and diesel). Within the present context of economic crisis and controversy about the role of biofuels, these figures might be revised.

The USA announced an extremely ambitious objective of 30% biofuels in the transport sector in 2030.

Ethanol is by far the most extensively used biofuel worldwide. It is produced by fermentation of sugar, obtained from plants such as sugar cane and sugar beet. Production of ethanol from sugar cane is widespread in Brazil, a country which has considerably developed the use of ethanol as a fuel. It can also be produced from starch derived from cereals such as maize and wheat. The sugar required for the fermentation step is obtained beforehand from starch under the effect of an enzyme. The production of ethanol from cereals, in particular maize, is mainly carried out in the USA, which is the second major ethanol producer.

Ethanol can be used in gasoline engines. In Europe, however, it is widely used as ETBE, a compound formed with isobutene, so that it can be more easily incorporated into gasoline. In the future, direct use of ethanol as fuel is likely to take off with the use of *flex-fuel* engines, already widespread in Brazil, the USA and Sweden, and launched officially in France at the beginning of 2007: flex-fuel vehicles can run on traditional gasoline, E85 fuel containing 85% ethanol (superethanol), or a mixture of the two in any proportions after filling up in different service stations.

The global production of ethanol reached 40 Mt in 2006, including 75% for applications as fuel, the largest share of the production being concentrated in Brazil and the USA. Its share in energy terms represents more than 80% of the biofuels market (83% in 2006).

In Europe, **vegetable oil methyl ester** (VOME), obtained from rape seed oil or sunflower oil, is currently the most widely used biofuel, as it can be incorporated in diesel fuel, for which the demand in the European Union is higher than for gasoline.

Up to 5% VOME can be incorporated in diesel fuel without the need for any significant engine modifications. Higher incorporation levels are possible, but the engine must be modified accordingly.

An area equivalent to 30–40% of the current agricultural land [68], whether in Europe or the USA, would have to be dedicated to biofuel if production is to reach a level equivalent to 10% of fuel consumption, which seems neither possible nor acceptable.

The competition for land use with crops needed for food production has generated much debate about biofuels. It is therefore essential to be able to produce biofuels from a feedstock which cannot be used for food applications.

Since it is not processed for human food, the use of lignocellulosic biomass (wood, agricultural waste such as straw from cereals and oleaginous plants, fast rotation crops on agricultural areas, etc.) as raw material would considerably increase the biofuel production potential.

In order to produce liquid fuels from lignocellulosic biomass, we will nevertheless need to develop 'second generation' technologies, which have not yet been proven either industrially or economically. Two main conversion pathways are used:

- The *thermochemical pathway*, under development, consists of converting biomass under the effect of heat into a gaseous or liquid phase. Liquid fuels can be obtained through gasification followed by Fischer-Tropsch synthesis. The BTL (biomass to liquids) process is similar to GTL (gas to liquids) and CTL (coal to liquids) processes used to produce synthetic fuels from natural gas or coal, which are described in Chapter 7. It offers the advantage of producing high-quality liquid fuels, in particular diesel, which can be used directly in current engines.

    Pyrolysis of lignocellulosic biomass in an oxygen-free atmosphere produces a liquid phase called 'bio-oil'. This product is rich in oxygenated compounds, making it immiscible with hydrocarbons. It can be converted into liquid fuels by gasification. Its conversion into fuel by hydrogenation is also one of the research pathways being investigated.

- The *biochemical pathway* requires fractionation of the lignocellulosic biomass into its three fractions: cellulose, hemicelluloses and lignin. Glucose can be obtained from cellulose under the effect of enzymes. The glucose then undergoes a fermentation step to obtain ethanol. The aim of current studies is to improve the performance of this conversion process. The research work also concerns the conversion of hemicelluloses to obtain other sugars, the pentoses, whose subsequent conversion into ethanol by fermentation is also being studied. This conversion, which is more difficult than the case of the cellulose fraction, has not reached the industrial stage [67]. Lignin is separated and can be used as a fuel to supply energy. Production of cellulosic ethanol is still currently more expensive than production of ethanol from other sugar plants. In view of the extensive research and development work in progress we can expect production to become more competitive.

Large-scale development of biofuels is only acceptable if compatible with sustainable development criteria. The risks associated with the production of biomass intended for use as biofuels must be carefully analysed (competition with food uses, consumption of water, fertilisers and pesticides).

If we restrict ourselves to the current first generation processes, competition with food uses appears when the incorporated rate in fuels exceeds 5–10%. The new processes, whose production is based on lignocellulosic biomass, must therefore be implemented to reach higher penetration levels.

Care must be taken in this case to ensure that the biomass is produced under conditions acceptable for the environment and, in particular, does not result in deforestation or irreversible soil degradation.

Improvement of the CO_2 balance, a critical selection criterion, must be assessed by performing a complete analysis of the life cycle, from the biomass production step through to use of the biofuel in the engine. It is absolutely essential to perform this life cycle analysis, since some cereal-based ethanol production processes, requiring large quantities of fossil energies to convert the biomass, offer little or even zero benefit.

In contrast, the more recent lignocellulosic biomass conversion pathways currently being explored reduce the global CO_2 emissions by 70–90% over the entire biofuel production and utilisation cycle.

Emissions of greenhouse gases other than CO_2, in particular nitrous oxide (N_2O), resulting from the use of nitrogen-containing fertilisers, must also be taken into account.

In order to achieve high penetration rates of more than 10%, while respecting sustainable development criteria and without compromising the production of biomass for food use, new biofuel production processes based on lignocellulosic biomass must be developed.

These criteria must always be respected, no matter which process is implemented. This is an essential condition if biofuels are to make a significant contribution to solving energy and climate change problems.

The role of hydrogen

Apart from the special case of biomass, renewable energies can only be used in the form of electricity (wind, photovoltaic) or possibly low-thermal-level heat (thermal solar and geothermal).

Energy diversification will therefore involve first increasing the use of electricity (in particular through heat pumps) and, secondly, introducing an energy vector which can serve as fuel and supply high-thermal-level heat, especially for industrial applications, without emitting CO_2.

Hydrogen is capable of playing this role, provided that a number of major obstacles can be overcome (cost, the need to set up transport and distribution infrastructures, storage). Hydrogen can in fact be used to store energy, but to reach sufficient storage densities, relatively expensive devices must be implemented: storage under very high pressure (600 bar), liquid at very low temperature or solid ('hydride' type inclusion compounds).

Hydrogen is also the most suitable fuel for fuel cells. Fuel cells convert the chemical energy of combustion directly into electricity with very high conversion efficiency, which is why they are often considered as the energy converter of the future. A distinction is made between the low-temperature cells, including the proton exchange membrane fuel cell (PEMFC) and its variants based on methanol or ethanol (DMFC and DEFC), and cells operating at a higher temperature, the alkaline fuel cell (AFC), molten carbonate fuel cell (MCFC) and solid oxide fuel cell (SOFC) [71]. Despite significant progress, the fuel cells, in particular the low-temperature fuel cells, which are intended for large-scale distribution, are still faced with lifetime and cost problems. The high cost is due in particular to the use of a platinum catalyst.

For the time being, therefore, their application is restricted to niche markets (military applications, emergency generator sets and portable electronics, developing market for miniature cells, hydrogen PEMFC or methanol DMFC).

Hydrogen can either be produced from fossil fuels or by electrolysis:

- Initially, hydrogen will be produced from natural gas, coal or biomass, via a synthesis gas. Synthesis gas is obtained from natural gas by an endothermic steam reforming reaction, in the presence of steam and from coal or biomass by an exothermic partial oxidation reaction in the presence of oxygen[4]. The synthesis gas is then converted in the presence of steam into a mixture of hydrogen and carbon dioxide ('shift conversion' reaction).

  Provided that the CO_2 is captured and stored, the hydrogen obtained can be used to produce electricity from coal or possibly natural gas without CO_2 emission.

  Consequently, production of hydrogen from fossil fuels is only of any interest if combined with CO_2 capture and storage. Centralised production facilities are therefore preferred, since CO_2 capture for

---

[4] Exothermic reaction: reaction occurring with production of heat.

storage in the subsoil is difficult with small-capacity decentralised facilities.

Hydrogen produced with CO_2 capture and storage, being of neutral carbon content, can also be used for petroleum refining and chemistry applications, in particular to obtain petroleum fuels of better environmental quality.

In addition, hydrogen (and therefore energy) produced with CO_2 capture and storage from biomass has negative carbon content, with a net removal of CO_2 from the atmosphere, as indicated previously.

• Although hydrogen produced by electrolysis, from nuclear or renewable origin electricity, is currently more expensive than hydrogen produced from natural gas, it should eventually become cost-effective. There seems little point in converting it directly into electricity, since it is simpler and cheaper to transport electricity as it is. Provided that the electricity is obtained from nuclear or renewable sources, it can be used in industry to improve the carbon balance. It can also be used as fuel on board vehicles.

Use of hydrogen as a means of storage can be used to offset the intermittent nature of the renewable energy supply or to modulate the supply of electricity from base-load production, i.e. at constant power, which corresponds to the conditions required with nuclear energy. In this case, a step to produce hydrogen by electrolysis is necessary, followed by a step to generate electricity in a fuel cell (or in a combined cycle). These successive conversions increase the costs and decrease the overall efficiency.

With on-board storage, use of hydrogen as a fuel in a vehicle offers the advantage of potentially avoiding CO_2 emissions while, at the same time, diversifying the energy sources. The progress made several years ago in the field of fuel cells raised hopes that a propulsion system consisting of a fuel cell, powered by hydrogen and producing the electricity necessary to drive the electric propulsion motors could represent the ideal solution sought in the field of transport. These hopes are from being realised, however. Numerous difficulties remain, concerning both use of hydrogen and development of the fuel cells (cost, reliability). The storage system is also a difficult issue. Hydrogen can be stored in gaseous form at high pressure (700 bar).

Hydrogen can also be stored in liquid phase, but at low temperature, resulting in high costs and making extended storage difficult. Studies are being conducted on alternative forms of storage, in particular as hydrides, but no competitive solution has yet been found. With the current systems implementing compressed hydrogen, a storage density of 4% by weight can be reached, expressing the weight of hydrogen stored against the total weight of the storage system. It has been estimated that this value would have to be approximately doubled in order to fit a storage system carrying the 8 kg of hydrogen required to achieve autonomy of about 500 km on a vehicle equipped with a fuel cell [69]. Since the low heating value (LHV) of hydrogen is 120 MJ/kg, a storage density of 4% by weight corresponds to an energy storage density of 1.33 kWh/kg. This density is already much higher than that which could be obtained using electrochemical batteries, since the storage density of the best batteries does not currently exceed 70 Wh/kg, as we saw in Chapter 5.

Large-scale distribution of hydrogen for a range of decentralised uses and for transport, creating a 'hydrogen civilisation' [70], is likely to take several decades and therefore seems uncertain.

The fact still remains that the energy transition will require the implementation of new energy vectors and new energy storage means. From this perspective, the use of hydrogen is one of the avenues to be explored.

Hybridisation of sources

The energy transition will be favoured by setting up hybrid systems, powered by different energy sources.

- *In the field of transport*, plug-in hybrid propulsion systems will play this role.

  The autonomy of the current hybrid vehicles, for example the Toyota Prius, is extremely limited (about 2 km) in purely electric mode. In addition, all the energy consumed is produced from the on-board fuel.

  Increasing the capacity of the electricity storage system and making the electric battery rechargeable will offer far greater possibilities, such as most urban transport requirements. The petroleum fuel consumption of this type of plug-in hybrid vehicle (PHV) can therefore be considerably reduced. In France, 60% of the daily trips would be covered by a PHV with an electrical autonomy of 30 km (PHV 30), this proportion reaching 70% for an electric autonomy of 40 km (PHV 40). Long-distance trips in France represent only 30% of the annual mileage.

  With a PHV offering autonomy of 60 km in purely electric mode (PHV 60), the average emission level would be about 40 g CO_2/km, a figure four times less than current emission levels. This assumes, however, that the electricity consumed is itself decarbonised.

By using a thermal engine which can run on a biofuel such as ethanol produced from lignocellulosic biomass, the dependence on oil and the CO_2 emissions can be reduced even further [74].

– *In electricity production*, the future electricity grid will be able to optimise the use of various types of centralised or decentralised generators, associated with variable capacity storage systems.

A global electricity distribution network could eventually manage various types of production system (centralised or decentralised, intermittent or continuous) as well as various types of demand (concentrated or dispersed, continuous or intermittent). It must be equipped with energy storage devices offering sufficient flexibility as well as protection systems to improve the network stability, such as FACTS (flexible alternative transmission systems), which include, in particular, systems whose reactance is controlled according to the state of the network [75].

A global network such as this has been compared with the Internet and the term 'Electranet' coined to emphasise the similarity.

This type of network forms a hybrid system connecting users to different energy sources.

It represents one of the means to be implemented to simplify a transition during which the share of fossil energies will be progressively reduced.

The development of 'smart grids' is one of the promising innovations to be explored.

Research actions required

... **For reducing the energy carbon content**

- **Nuclear**
  - Development of fourth generation reactors
  - Waste treatment and storage
  - Development of fusion reactors

- **Renewable energies (wind, solar, geothermal)**
  - Development of offshore wind turbines
  - Improvement of photovoltaic systems
  - Deep geothermal energy

- **Production of energy from biomass**
  - Treatment of combustion gases
  - Gasification
  - Production of biogas
  - Production of electricity and/or heat with CO_2 capture and storage
  - Biofuels
  - Improvement of the first generation technologies
  - Development of processes for the production of fuels from lignocellulosic biomass by fermentation and thermochemical pathway
  - Direct liquefaction of biomass
  - Production of biofuels with negative carbon content (production with CO_2 capture and storage)

- **Hydrogen**
  - Synthesis gas production processes
  - Development of hydrogen production with CO_2 capture and storage
  - Production of hydrogen from biomass, including production of hydrogen with negative carbon content

- **Hybridisation of sources**
  - Development of 'smart grids'
  - Plug-in hybrid systems

7

Securing the Supplies of Fossil Fuels

The role of fossil fuels

While our aim is to reduce as rapidly as possible the share of fossil fuels in the world primary energy supply, substantially reducing the amount of energy consumed and deploying renewable energy sources will take time. What can the role of fossil fuels be during this transition period and how can they contribute to help in achieving a transition aimed at their disappearance?

Furthermore, is the objective to widen and diversify fossil fuel sources reconcilable with the need to reduce CO_2 emissions? Considering only coal, its proven reserves presently amount to around 900 billion tonnes, which means that 3700 billion tonnes of CO_2 will be produced, if this coal is consumed. This is much more than the total emissions of carbon which might be acceptable during the next fifty years, if we want to keep to the objective of a mean temperature increase below 2 °C.

Three reasons have to be taken into account for maintaining substantial efforts in the area of fossil fuels:

- As has already been mentioned, fossil fuels will keep a significant share of the primary energy supply during the whole transition period. Even if their consumption is greatly reduced, it will be necessary to satisfy the demand, without any sudden breakdown in supplies. The problem to solve is not related to the absolute level of

Energy and Climate: How to achieve a successful energy transition   Alexandre Rojey
Copyright © 2009 Society of Chemical Industry

reserves, but rather to the production level to be maintained during the whole transition period.

– It is necessary to diversify supply sources, in order to reduce the risks and especially the geopolitical risks which become critical if the number of suppliers is small.

– All fossil fuel energy sources are not equivalent. Natural gas leads to CO_2 emissions per unit energy produced which are much smaller than those which result from the use of coal.

It would be therefore very detrimental if a short term tension on hydrocarbon supply leads to a rapid and massive shift towards coal, especially before the technology of CO_2 capture and storage becomes operational.

To secure the supply of fossil fuels, a first imperative condition is to reduce their consumption, as discussed in Chapter 5. It is also important to avoid a breakdown of the supplies during the transition period. Finally, it is necessary to diversify the supply sources and to favour those which have the smallest impact on the environment.

The end of abundant and cheap oil

The first oil crisis in 1973 and then the second in 1979 have had the effect of warning signals, favouring the development of 'technological' oil, as illustrated by the tremendous progress observed in the area of offshore oil production.

The evolution of oil prices is shown in Figure 7.1; the price of oil per barrel is expressed both in current US $ and in constant US $ (upper curve). After a long period of very low prices, which followed the first discoveries, the history of oil has become much more turbulent over the last thirty years. In 1973, the first oil crisis resulted from the Yom Kippur war. Then, in 1979, the second crisis followed the Iranian revolution and the Iran–Iraq war. A counter crisis favoured by the development of technological oil and a weakening of the role played by OPEC followed, leading to a new drop in oil prices. It occurred during a period which was marked by the end of the Cold War and the easing of international relations.

Within this context, there was a feeling that the difficulties were over and that plenty of hydrocarbon resources were available. Low prices also had a negative impact on investments concerning technological oil and alternative energy sources.

A revival of political tension, the war in Iraq which followed the Kuwait invasion, the spreading of terrorism, the increase in world demand and

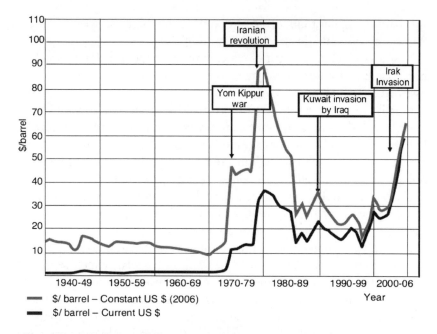

Figure 7.1   Evolution of the price of crude oil (*Source*: BP Statistical Review)

a more pessimistic outlook concerning oil reserves led in 2007 and the beginning of 2008 to a sharp increase in the price of oil which reached 100 US $/barrel and then 140 US $/barrel.

The financial crisis of autumn 2008 provoked a new sharp drop in price. Due to these very rapid oscillations, the price of oil tends to become unpredictable in the short term. In the longer term, it appears that the era of cheap and abundant oil is finished and that a general trend towards increasing prices should be observed in the future.

Oil reserves: the present situation

Hydrocarbon reserves are constituted by oil and gas contained in identified fields, which can be produced under economically and commercially acceptable conditions [76].

They take into account the economic and technical conditions, such as the recovery yield which can be achieved with the available technologies, prevailing at a certain time.

According to the level of certainty with which these reserves are known, it is possible to make the distinction between reserves which are proven (probability higher than 90 %), probable (probability between 30 % and 90 %) and possible (between 10 % and 30 %).

Resources include quantities which have been identified, but which are not considered commercially exploitable and those which have been assessed as still to be discovered.

Finally, ultimate resources comprise all hydrocarbons present or estimated, including those quantities for which no possibility of extraction has been identified. Ultimate resources include therefore hydrocarbons which cannot be transformed into reserves by using any identified technology. This appraisal can of course change with time.

Unconventional resources are those which cannot be exploited by using standard practices (or 'conventional' means), and are constituted in the case of oil by extra-heavy oils, asphalt sands or oil shale.

With the present technical and economic conditions, proven reserves of oil are estimated at 162 Gt (i.e. 1200 billion barrels), which represents around 40 years of consumption at the present rate.

The diagram in Figure 7.2 represents the different oil resources. The oil production cumulated since the beginning of oil exploitation amounts to 137 Gt. Present proven reserves, with an average oil yield recovery factor of 35%, represent 162 Gt. Fields which remain to be discovered (according to a probabilistic approach based upon the investigation of all the

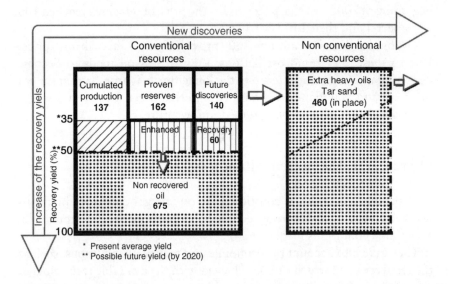

Figure 7.2   Oil reserves and resources (*Source*: IFP). Reproduced with permission from *Energie & Climat: Réussir la transition énergétique* by Alexandre Rojey, Éditions Technip, Paris, 2008

sedimentary basins of the planet), if they were exploited with the same recovery yield, represent a further 140 Gt.

Increasing the recovery yield from 35 to 50% (through the use of new technologies) would make it possible to increase present proven reserves by 65 Gt and resources resulting from future discoveries by 60 Gt.

With these assumptions, a total amount of 265 Gt (or 1850 billion barrels), might therefore be added to the present proven oil reserves.

The resources of nonconventional crude oil are huge and amount to 460 Gt, but for the time being only a comparatively small share of these resources may be exploited.

These values can seem reassuring but two important points should not be forgotten. First, the world consumption rate has considerably increased during recent years: more oil has been consumed since 1980 (a period of 26 years) than during the whole previous period (more than a century!). From the supply point of view, the amounts of oil resulting from new discoveries do not correspond to the consumption rate by a long way. Very few giant fields have been discovered in recent years, whereas presently exploited giant fields deliver half of the total world production. Thus, the discovery of the four biggest fields each supplying more than 1 million barrels per year occurred before the 1980s.

Secondly, large uncertainties exist concerning the exact level of proven reserves. The values which are published depend upon the declarations of producing countries which may have an interest in overestimating the level of their reserves.

The uneven geographical distribution of oil reserves creates a geopolitical risk for the security of supplies, which is reinforced by the growing dependence of consuming countries.

Towards a more technological oil

An oil field can be exploited only during a limited period of time. The production of an oil field starts by increasing, then reaches a production plateau and finally declines over a period of time which can be of variable length.

The moment when the maximum of the world oil production might be reached ('peak oil') is often debated and has been frequently highlighted as an imminent risk by ASPO (*Association for the Study of Peak Oil & Gas*).

By considering the evolution curve of the quantity of oil discovered over the past years, the American geologist King Hubbert was able to show that production followed the same curve with a time shift of 33 years.

As the peak of discoveries occurred in the USA in 1937, he predicted in 1956 that the production peak would occur in 1970, that is to say fourteen years later, which is what effectively happened [78].

Since that time, numerous attempts have been made to apply this method on a world scale level, resulting in many predictions concerning the date of the future peak oil [82]. However, this method, which was very successfully used for predicting the peak oil in the USA, is difficult to apply on a world scale level.

The date when the world production of oil will reach a ceiling is not only a function of the absolute level of resources, but also of technical and financial means which can be used for exploiting them. It must be stressed that the concept of peak oil is much more complex on a world scale than on a regional scale. The increase of prices which will result from such a situation will lead to an increase in production from nonconventional sources, which are not presently taken into account.

Different predictions concerning a more or less imminent peak oil have been published recently.

ASPO estimates that peak oil will occur towards 2010, with a production ceiling around 90 million barrels per day [81]. The Douglas-Westwood Company predicts peak oil in 2016 [81] and the oil company Total, in a report published in 2003, predicts about 2025 [83].

Taking into account the large fluctuations observed concerning the worldwide production of crude oil, which depend upon the economic conditions observed at a given time, the exact date of the production 'peak' represents a somewhat theoretical issue.

At the approach of the maximum of production, the oil market is likely to become very unstable and rather than a peak, a strongly undulating plateau might be observed.

This is perhaps what is already happening but very unstable economic conditions would make it difficult to observe such a peak even when it happens effectively.

It seems likely that a maximum level of production, at around 95 billion barrels per year, will be reached within the next twenty to thirty years. Since world production is already about 87.5 billion barrels per day, the potential increase of this production level is fairly limited.

The growth in oil demand will be the key factor. If it continues to grow at the present rate, it will be difficult to ensure a supply at the corresponding level. The main problem stems from the necessity to ensure the investments needed to maintain the required production level. There is a risk of a gap between supply and demand that no other energy source is able to compensate rapidly.

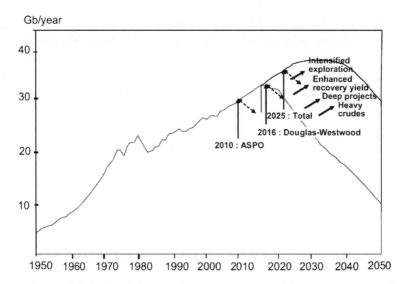

Figure 7.3   Pushing forward the envelope of oil production (*Source:* IFP). Reproduced with permission from *Energie & Climat: Réussir la transition énergétique* by Alexandre Rojey, Éditions Technip, Paris, 2008

In order to avoid a crisis in the comparatively short term, it is therefore necessary to push forward the limits of oil production.

For maintaining world production at the required level over several decades, it will be necessary to discover new fields and to exploit the existing fields in an optimal way in order to maintain the life of fields as long as possible. 'Technological' oil is going to play a growing role in the years to come.

As can be seen from the diagram in Figure 7.3, an increased use of high performance technologies makes it possible to push forward the limits of oil production by around twenty years. These technologies contribute by:

- Making possible new discoveries, through the use of more and more sophisticated exploration technologies, such as elaborate seismic imaging methods of the subsurface.
- Improving the recovery yield, by applying on a large scale enhanced oil recovery methods: water and gas injection (secondary recovery), injection of CO_2, steam, polymers or surfactants (tertiary recovery).
- Increasing the productivity of oil wells; technologies currently exist for drilling deviated, horizontal or even 'fish bone' well architectures, in order to improve drainage of oil deposits.
- Developing offshore oil production, including deep offshore and areas which have not been widely explored until now, such as the

Arctic zone. The exploration of new areas has to be undertaken carefully, in order to protect the environment. Progress in the area of offshore production has been spectacular: oil can be produced now at water depths reaching or even exceeding 2000 m. Further progress is still expected; it is estimated that 40 % of deep offshore reserves still to be discovered are located between 2000 m and 3000 m, and 30 % are between 3000 m and 4000 m water depth. The global volume of the reserves still to be discovered in the deep offshore and in the Arctic zone is estimated at around 100 Gb [77].

- Putting into production nonconventional oil fields (heavy and extra-heavy oils). The future potential oil production which might result from the exploitation of these fields is very significant and the future impact of these prospects will be presented below.

These different means will be effectively deployed only if the oil price remains high enough over a long period and if the required investments are provided, involving the use of the required advanced technologies. The further delay which might result from the use of these innovative technologies should help to manage the transition. It should not be used for postponing the other actions which need to be undertaken. On the contrary it must be considered as an opportunity for deploying them actively.

Extra-heavy oils and shales

Extra-heavy oils are formed by deposits at a comparatively shallow depth, more or less deeply degraded (oxidation by bacteria), of high specific gravity and viscosity.

They form either underground layers or tar sands emerging at ground level. A fraction of these heavy oils can be produced directly, when the oil reservoir is located deep enough to reach a temperature which makes the oil fluid enough to circulate in the production well as a liquid phase. This is the case of the heavy oils presently exploited in Venezuela. If it is not the case, these heavy oils can be exploited by surface mining or by drilling, using a thermal process to reduce the oil viscosity. The method most often used is to inject vapour to increase temperature and so reduce viscosity. Thus, the SAGD (steam assisted gravity drainage) process involves injecting steam in a horizontal well and discharging the heated oil through a second horizontal well located above the injection well [79]. Such processes are already used, and production costs, which can be below 20 US $

per barrel, are economically acceptable. Still, these processes require a great amount of energy, generate very significant emissions of CO_2 and require large quantities of steam and therefore of fresh water. To exploit extra-heavy oils in an acceptable way, it will be necessary to develop technologies which minimise the impact upon the environment and especially CO_2 emissions.

Canada and Venezuela presently produce 1.8 million barrels a day, i.e. 2 % of the world oil production. In Canada alone, the production of heavy oils from tar sands exploited by surface mining could reach 1 million barrels a day by 2010 and 5 million barrels a day by 2030.

In terms of resources, extra-heavy oils represent a huge potential, around 1200 Gb each for Venezuela and for Canada. It is estimated that around a quarter of these resources, i.e. 600 Gb, can be exploited by using already known technologies [80].

An oil shale is formed by a fine-grained sedimentary rock containing an organic matter which has not been completely converted into oil. Shale oil can be exploited by using mining technologies, but its exploitation has a strong environmental impact: in the case of surface mining, destruction of vast natural areas, increased erosion, large energy consumption for processing the shale and extracting the oil, acid drainage, and air pollution. To date, its exploitation has not appeared to be economically competitive, as compared for instance with coal mining.

Natural gas

The development of natural gas has been quite rapid in recent years. It has become a widely used and excellent fuel for power plants, presenting strong environmental advantages (providing a clean fuel emitting 50 % less CO_2 per unit energy output than coal and 30 % less than fuel oil) which can be used in an efficient and flexible way. Natural gas fired power plants require comparatively low investments and are quick to install. Combined cycle power plants, which associate a gas turbine with a steam cycle exploiting heat recovered from the gas turbine exhaust gases, reach a high efficiency, close to 60 %. Gas fired power plants emit few pollutants to the atmosphere. Compact, quick to build and requiring a comparatively small investment, they have developed rapidly.

Natural gas is used also for residential heating applications and as a fuel in industry. Stored under pressure in a tank, it can even be used as an engine fuel, called NGV (natural gas for vehicles). This option can be applied for diversifying engine fuel supply sources. If it is properly used in a well adapted engine, it can lead to excellent results from the environmental standpoint, both in terms of a low pollutants emission at the local level and CO_2 emissions.

The growing demand for natural gas combined with a growing distance between reserves and consuming areas led to rapid progress in international trade, increasing at an average rate of 6.8 % per year over the last twenty years, reaching 845 billion m^3 in 2005 [84].

Transportation by pipe requires significant infrastructures, which leads to a mutual dependence between the supplier and the consumer. This mutual dependence represents a stability factor, but can also be considered as a threat within an uncertain geopolitical situation.

In recent years, LNG (liquefied natural gas) transportation by tanker has grown rapidly. As a result of the cost reductions already achieved at all the stages of this transportation chain and the increased flexibility it brings compared with pipe transportation, LNG represents a major growing factor for the international trade of natural gas.

LNG transportation by tanker is expected to grow at a rate of 7 % per year until 2020. The level of the international LNG trade has reached around 515 billion m^3 ensuring 38 % of the international gas transit as compared with 22.3 % in 2005 [85]. The development of the international LNG trade therefore represents an important factor for diversifying gas supplies.

When the distance between the production and the consumer sites becomes too large, the transportation cost can become prohibitive and prevent the exploitation of some gas fields. The production of synthetic fuels from natural gas is a new outlet for natural gas and can be considered as an alternative to the direct transportation of natural gas in such cases. The future outlook in this area is discussed later in the chapter.

The recent increase in the price of natural gas, which remains strongly correlated to the price of oil, and also the economic crisis which occurred in the second half of 2008 have made the future rate of growth more uncertain. Most recent estimations were based upon an average growth rate of around 2 % for the overall gas demand during the next two decades [85].

The proven reserves of natural gas amount to 181 000 billion m^3, representing around 64 years of consumption. Taking into account that 1000 m^3 of gas equals 0.9 toe, the reserves represent therefore around 163 billion toe, i.e. an amount close to the level of the oil reserves.

Although this situation is more favourable than in the case of oil (64 years of reserves at the present rate of consumption as compared with 40 years), the growth of the consumption could lead, if present reserves are not sufficiently renewed, to a gas peak by 2050.

Contrary to what happens in the case of oil, the recovery yield in the case of natural gas is comparatively high and therefore cannot be increased substantially.

However, natural gas has been explored less intensively than oil and the reserves of natural gas are therefore probably underestimated. There remains a large exploration potential for discoveries to be made. The renewal of gas reserves will be achieved in the future mainly through new discoveries and the exploitation of unconventional resources.

Resources of unconventional gas are very large. These resources include deep deposits, contained in low permeability sedimentary rocks ('tight gas'). Coal bed methane is already widely exploited in the USA and its production might be greatly increased in other regions of the world.

Very large quantities of methane are also trapped in the form of hydrates. Hydrates are solid inclusion compounds, which can be formed under pressure and at a sufficiently low temperature. Water molecules form a lattice in which molecules of light hydrocarbons, such as methane, can be trapped. They form deposits in various cold regions of the world, near the Arctic zone, in Siberia and in Canada. Hydrates are also trapped under the bottom of the oceans as a result of the huge pressure due to large water depths. For the time being, the exploitation of hydrates remains very limited, due to many difficulties resulting from the need to handle a solid phase, and also due to the fact that hydrates are often present in a very dispersed way.

In the future, an increase in energy prices could help to increase exploration efforts, giving access to new reserves of natural gas. It is therefore premature to consider a short term decline of natural gas production, due to a depletion of resources.

The advantages of natural gas in environmental terms are quite good, both at the local and the global level, with a level of CO_2 emissions per unit of energy output much lower than in the case of coal. It is therefore an energy well adapted to the transition period, with a much lower environmental impact than coal, especially before the effective deployment of CO_2 capture and storage technologies.

The comeback of coal

Due to the size of coal reserves and a distribution more favourable for large countries with high consumption than in the case of hydrocarbons, the development of power generation from coal is expected in the world, especially in the USA, India and China [86].

Coal reserves are huge: proven reserves amount to 900 Gt, representing 160 years of consumption at the present rate. Coal resources reach 3700 Gt. At the present time, the level of coal reserves does not appear as a limiting factor for the economic development of coal. Nevertheless, it is necessary to take into account the fact that the information concerning coal reserves is not always reliable. Some countries, such as China, have not revised the level of their current reserves for many years, despite the very substantial amounts of coal already produced.

Different types of coal have been formed through geological processes which occurred in the past. The most mature and best quality types of coal are bituminous coal and anthracite. More recently formed types of coal include lignite and peat. Their heating value is lower and their production is concentrated in countries where they can be exploited at low cost using surface mining. As it is the best quality coal which is preferably exploited, its share in the world reserves tends to decrease. The production of coal in Europe, which most usually requires deep mining, has strongly declined.

The recent growth in the price of oil, which resulted in a high price for natural gas, favoured the use of coal which remained significantly cheaper, as shown in Figure 7.4.

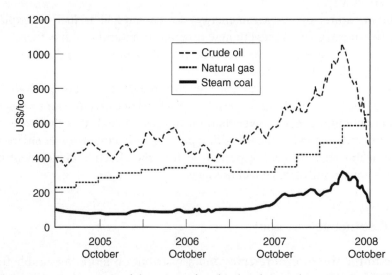

Figure 7.4 Comparison of the prices of coal, oil and natural gas (*Source*: ATICS)

Even so, the price of coal also increased sharply and rapidly in 2008. The CIF price of steam coal reached around 300 US $/toe in July, falling back to around 200 US $/toe in September 2008 (compared with 600 US $/toe for natural gas and 700 US $/toe for oil at the same time).

In China, the rapid increase in demand has led to a very ambitious planning of renewal and growth of the power generation capacity in the years to come. China plans around 1 GW of additional capacity each week over 20 years. The additional capacity in China amounted to 50 GW in 2004, 70 GW in 2005 and 102 GW in 2006. China aims to install an overall capacity of power generation between 1200 GW and 1300 GW by 2030 [87].

The world consumption of steam coal for power generation might therefore increase from 1500 Mtoe in 2000 to 2500 Mtoe in 2030. Coal ensures around 40 % of power generation, which amounts to around two-thirds of the worldwide coal consumption.

Furthermore, if, in the future, it becomes difficult to satisfy the demand for engine fuels by deriving them only from oil supplies, the production of synthetic liquid fuels from coal will increase rapidly. The yield of such a transformation is unfortunately comparatively low: an amount of 2.5 toe of coal is needed to replace 1 toe of oil. Large emissions of CO_2 therefore result from such a transformation. Synthetic fuels can also be produced from natural gas or biomass and the future outlook for these synthetic fuels is discussed in the next section.

The increase of coal production poses difficult problems from the environmental standpoint, both at the local level (emissions of different contaminants: SO_x, NO_x, solid particles, mercury, etc.) and at the global level, due to the level of CO_2 emissions. The situation is especially worrying in China, where coal accounts for 77 % of power generation and in India where 70 % of power capacity is coal based.

Technologies for 'clean coal' power generation are available or under development. The main difficulty stems from the need to limit CO_2 emissions. If the comeback of coal is confirmed, it will cause a significant increase in CO_2 emissions. CO_2 emissions per kWh of electricity produced from coal are from two to three times higher than in the case of natural gas, as a result of the combination of two factors (a lower efficiency and a higher level of CO_2 emissions per output energy unit) [88].

The situation can be improved by increasing the efficiency of coal-fired power plants. The efficiency of the most recent coal-fired power plants is close to 45 % and in the future efficiency close to 50 % is expected, which is well above the world average.

However, in order to limit the risks of climate change, the development of the use of coal for generating electricity or for producing synthetic fuels is acceptable only if the CO_2 emitted is captured and stored underground.

The quantities of CO_2 which might result from the combustion of all the carbon trapped underground in coal deposits are much too large to be released into the atmosphere [89].

Therefore, although coal certainly has a role to play in the energy mix, as long as CO_2 capture and storage technologies are not effectively implemented, its rapid development represents a threat from the standpoint of climate change.

The deployment of these technologies, which are presented in the next chapter, will require time and meanwhile the development of power generation (as well as the production of synthetic fuels) from coal should be considered with care.

Synthetic fuels

Liquid fuels are presently almost totally produced from crude oils through refining processes. This dependence on oil might be reduced in the future though the use of synthetic fuels and biofuels.

Different types of feedstock can be used for producing synthetic fuels (natural gas, coal and biomass). Liquid fuels are produced through a synthesis process from a gaseous mixture of carbon monoxide (CO) and hydrogen; this 'synthesis gas' is generated by gasifying the initial raw material through a thermochemical process.

The most favoured pathway is presently the Fischer-Tropsch synthesis, so named from the two German chemists, Franz Fischer and Hans Tropsch, who developed this process during the Second World War, in order to reduce the dependence of Germany on imported oil. The long chain hydrocarbons which are produced by the Fischer-Tropsch process can be transformed into fuels of excellent quality (gas-oil for diesel engines and kerosene for aviation). Synthetic fuels have been also produced on a large scale in South Africa, as an answer to the trade embargo which had been placed at that time. The units which were built at that time used coal as a feedstock. In 1991, South Africa also built a unit producing synthetic fuels from natural gas and presently South Africa is the biggest producer of synthetic fuels, with a production capacity of around 200 000 barrels per day.

Other players have also developed technologies in this area and the production of liquid fuels from natural gas (Gas To Liquids or GTL pathway) appeared as an attractive option for utilising large quantities of natural gas located far from consuming areas (stranded gas). In 2000, large projects were announced in Qatar, for producing liquid fuels from natural gas extracted from the North Dome field, which is the biggest gas field in the world. The Oryx GTL project has been initiated with the participation of Qatar Petroleum, of the South African company Sasol and the American company Chevron. A first 34 000 barrels per day (around 1.7 Mt/year) unit has been built and started production in 2006. Other projects have been announced. If they are implemented, the resulting capacity would exceed 800 000 barrels per day for Qatar alone. The investments required are huge: more than 900 million US $ for the 34 000 barrels per day unit built by Oryx GTL and 1.3 billion US $ for a comparable unit built in Nigeria.

Increasing costs for materials and manpower combined with the recent surge of the price of natural gas have contributed to a slowdown in operations. Nevertheless, the growth in demand for engine fuels should stimulate new investments in this area in the future.

In the longer term, if the price of oil rises and remains high long enough, the production of synthetic fuels from coal (Coal to Liquids or CTL pathway) can be expected to increase also. China has already decided to move in that direction. The biggest mining company in China, Shenhua, is planning to produce 30 million tons a year (around 630 000 barrels per day) of synthetic fuels from coal by 2020.

Other projects are considered, and if all these projects are implemented, China would produce within 15 years more than 70 million tons of synthetic fuels starting from 210 million tons of coal [87].

The production of synthetic fuels leads to comparatively large emissions of CO_2, especially if coal is the feedstock. In the case of the CTL pathway, around half of the carbon used is emitted to the atmosphere in the form of CO_2, while the other half remains contained in the liquid fuel which is produced. In the future industrial units producing synthetic fuels might become a source of very large CO_2 emissions, unless the CO_2 released is trapped and stored underground. Such an option can be considered for the very large units which are planned, but it will increase costs, which are already high.

Synthetic fuels can also be produced from biomass (biomass to liquids or BTL pathway; cf. Chapter 6).

Research actions required

... For securing the supplies of fossil fuels

- Exploration technologies
  - Basin modelling and geochemical methods
  - High resolution seismic methods
  - Other physical methods

- Improving the oil recovery yield
  - Modelling and optimising production
  - Improving production architectures
  - Secondary and tertiary recovery methods

- Offshore hydrocarbon production
  - Deep offshore production
  - Submarine production systems

- Extra-heavy crudes
  - Development of environmentally friendly processes
  - Hot production with low CO_2 emissions

- Natural gas
  - Exploration technologies for discovering new gas resources (deep gas, nonconventional gas)
  - Offshore liquefaction and chemical conversion
  - Unconventional gas (coal bed methane, hydrates, dissolved gas, tight gas)
  - New transport technologies (by pipe and tankers, compressed natural gas or CNG, mini LNG)

- Coal
  - 'Clean coal' technologies – flue gases treatment
  - Gasification technologies
  - High efficiency power plants
  - Electricity generation with CO_2 capture and storage (CCS)

- Synthetic fuels
  - Synthetic gas production processes
  - Production pathways from coal and natural gas
  - Production with CCS

8

CO$_2$ Capture and Storage

Stakes ahead

Fossil energy sources are still going to be used during the transition period. Therefore, it will be necessary to deploy CO$_2$ capture and storage (CCS) technologies on a large scale, especially with respect to the strong impact upon the environment of coal-fired power plants.

Climate change requires urgent actions. Unfortunately, the decrease of the share of fossil fuels in the world energy supply will not be rapid enough to reduce CO$_2$ emissions to the required level by 2050. For keeping the right overall balance, it will be necessary to capture and store at least a fraction of the CO$_2$ released by fossil fuels.

In order to avoid emitting into the atmosphere the CO$_2$ produced by combustion, different means can be considered:

- Producing biomass through a *photosynthesis* process, fixing the carbon and releasing oxygen into the atmosphere. Natural biomass reservoirs are called carbon *sinks*. In order to capture CO$_2$ and reduce its content in the atmosphere, it is necessary to grow *additional* quantities of biomass.
- Storing CO$_2$ deep underground: such a *geological storage* is the main option considered presently.
- Storing CO$_2$ in oceans: the risks which might result from such *ocean storage* (ocean acidification) are often considered too high to select such an option.

Energy and Climate: How to achieve a successful energy transition   Alexandre Rojey
Copyright © 2009 Society of Chemical Industry

- Reacting CO_2 with a basic rock, either after grinding the rock in order to obtain a mineral powder (*ex situ mineralization*), or within the geological formation itself (*in situ mineralization*).
- Producing a material or a chemical production through a reaction with CO_2. In such a case, the carbon is *recycled*.

These different options will be examined in this chapter. Whilst they seemed somewhat far off options a few years ago, at least some of them have now become an industrial reality.

Carbon sinks

In a broad sense, any natural reservoir able to absorb CO_2 is a **carbon sink**. The main natural carbon sinks are oceans and forests. Photosynthesis is the main carbon sequestration mechanism, but CO_2 can be also absorbed in the oceans by dissolution. Carbon dioxide exchanges between the atmosphere, vegetation and ocean surfaces are very large:

- 60 Gt of carbon is exchanged each year between vegetation and the atmosphere;
- 90 Gt of carbon is exchanged between ocean surfaces and the atmosphere;
- 40–50 Gt of carbon is exchanged between ocean surfaces and marine biomass.

The quantity of carbon emitted each year in the atmosphere as a result of fossil fuel consumption, i.e. 7 Gt/year, might appear small when compared with these exchanges. In fact, this additional quantity of carbon modifies a fragile balance, producing therefore a continuous shift in the CO_2 concentration in the atmosphere.

A first option for sequestering CO_2 consists of producing biomass to obtain a carbon sink, for instance by growing forests which will absorb carbon present in the atmosphere. It is necessary to grow biomass on large surface areas. Biomass thus produced can represent typically around 10 tons of carbon captured per hectare per year, amounting to 37 tons of CO_2. The implementation of such carbon sinks has become an economically viable option following the Kyoto Protocol, as carbon sinks can be taken into account in the carbon trade mechanisms.

Nevertheless, it must be underlined that the amounts of carbon thus sequestered are limited, as it is necessary to grow biomass over very large areas of land which, generally, are not available in a populated area.

Furthermore, it is necessary to ensure a rigorous management of the planted area, in order to achieve a positive carbon balance. Forests act as carbon sinks only during their growth period. A lack of maintenance or, even worse, deforestation at a later stage leads to a negative balance.

Biomass production has not only positive effects: agriculture and cattle breeding have a negative impact on the greenhouse effect due to CO$_2$ emissions resulting from the use of agricultural machinery, and also as a result of methane released by ruminants and NO$_x$ emissions resulting from the use of nitrate fertilisers.

CO$_2$ capture and transport

Carbon dioxide capture and geological storage is a way to implement artificial carbon sinks as it consists of storing large amounts of CO$_2$ underground.

This pathway has been extensively investigated worldwide during recent years, particularly in Europe, the USA and Japan. The basic principle is the following: CO$_2$ released from a large stationary source (industrial flue gases, CO$_2$ rich raw natural gas) is captured, concentrated, compressed and transported to an adequate geological formation in which it can be stored [90–92].

Carbon dioxide capture from combustion flue gases can be implemented through the use of technologies already available.

In most cases, solvent scrubbing processes are used, involving amines such as MEA (monoethanolamine), which are already used for natural gas processing.

The solvent absorbs CO$_2$ contained in flue gases by contacting the gases in an industrial column. Captured CO$_2$ is then released and separated from the solvent by heating in an industrial column. The solvent is then recycled and CO$_2$ is processed and sent to the storage location.

In the case of existing installations, flue gas scrubbing is preferably done at the 'post-combustion' stage, i.e. following energy production by combustion.

This option has the advantage of being readily applicable and does not require a full transformation of the existing installation if there is enough space area available, taking into account the area required by the equipment to be installed.

This equipment operates at rather unfavourable conditions using large volumes of flue gases at a low pressure and low CO_2 concentrations. The installations are bulky, costly and require fairly large amounts of energy, which can lead in some cases to almost double the energy consumption.

Therefore, in the case of new installations, other options have to be considered: one such option retrieves CO_2 before combustion by transforming the fuel which is used into a clean one through pre-combustion capture; another option consists of using pure oxygen instead of air in order to produce CO_2 concentrated flue gases (oxy-fuel combustion).

In the first case (pre-combustion capture), the fossil fuel is first converted into a synthesis gas, consisting of a mixture of carbon monoxide (CO) and hydrogen, either by steam-reforming of natural gas or by partial oxidation in the presence of pure oxygen from coal or biomass. The CO present in the mixture is then reacted with steam water during a 'shift-conversion' stage in order to produce a mixture of CO_2 and hydrogen.

Carbon dioxide is then separated from hydrogen at good operating conditions, i.e. at a comparatively high pressure and concentration, and sent to a storage location. Hydrogen can then be used for producing energy (electricity and/or heat) without any CO_2 emission.

This option has been considered for the FutureGen project launched in the USA. This project, for which a budget of around 1 billion US $ had been announced, was due to be demonstrated on the scale of a 275 MW power plant, planned to be operational by 2012. The project is now cancelled and restructured, which illustrates the difficulty of financing such demonstrations.

Oxy-fuel combustion is operated in the presence of pure oxygen. Carbon dioxide, which is at a high concentration in the flue gases, is easily separated from steam water with which it is mixed. This technology is already operational and industrial pilot plants are to be tested on a scale of around 30 thermal MW in Germany by the electricity producer Vattenfall and in France by Total.

In the case of the project led by Total, CO_2 captured in the Lacq setting will be transported through a pipe and injected in the depleted reservoir of Rousse, located 30 km from Lacq. Pure oxygen which is required for oxy-fuel combustion is currently produced by low-temperature distillation. New options are presently being investigated for transferring oxygen directly to the combustion zone. One of these options is to circulate a solid phase, which extracts oxygen from air and then releases it back to the combustion zone ('chemical looping').

After capture, CO$_2$ is generally transported in gaseous phase. Transport in a supercritical state[1] avoids any risk of a phase change. Carbon dioxide transport by tankers (road, rail or sea) is possible, but due to the large flow rates which are generally required, pipe transport is considered to be the most realistic option in the future. In the USA an extensive pipe network already exists for transporting CO$_2$ used for enhanced oil recovery (EOR).

Thus, CO$_2$ produced from natural underground reservoirs located in New Mexico and Colorado is transported to West Texas. Around 22 million tons of CO$_2$ per year are transported through a 3980 km long network of pipes to be used for EOR.

Transporting CO$_2$ by pipeline is therefore a well established technology and the deployment of CO$_2$ pipe networks does not raise any specific technical difficulty.

Furthermore, CO$_2$ is a stable gas, weakly toxic at least when diluted in the atmosphere, and nonexplosive. It is less dangerous to transport and to store than natural gas, for which the logistics of transport and distribution have been quite well mastered. Nevertheless, installing a pipe in a densely populated area often remains a difficult and expensive task.

Geological storage

The minimum period of time during which CO$_2$ needs to be stored is long but not unlimited. It has to last over one to two centuries, in order to cover the period during which the issue of CO$_2$ emission is likely to remain critical. Beyond such a period of time, CO$_2$ stored underground might be used as a source of carbon, and could be recycled by using a low carbon energy which should become widely available by that time. In order to keep the option fully safe, storage time durations which are presently considered amount generally to one or several thousand years.

Storage in deep oceans has also been considered as a potential option, but suffers two major drawbacks: firstly, the long term behaviour of such storage is difficult to model; secondly, the impact of an increase of the CO$_2$ concentration on marine ecosystems is difficult to assess and potentially dangerous. An accumulation of CO$_2$ might lead to an excessive acidification of the ocean. An increase of the temperature might

---

[1] Above a certain pressure, called the 'critical pressure' in the case of a pure component, it is no longer possible to obtain two phases, one liquid and one vapour, by lowering the temperature.

reduce the CO_2 solubility, leading to massive release of CO_2. For all these reasons, this option is not considered as acceptable by the European Union.

Underground geological storage is therefore the solution generally favoured. Numerous natural underground CO_2 reservoirs already exist in different regions of the world. Such storage is based upon the fact that sedimentary rocks are, in general, porous. Sandstone, which is formed by agglomerated sand particles, is a good example of such sedimentary rocks. The rock porosity is generally filled by water (salt aquifer). Carbon dioxide is injected through a well in the sedimentary rock and forces out the water. Using such a porous layer as an underground reservoir also requires the presence of a cap rock above the sedimentary layer, which acts as a seal preventing upward migration of CO_2.

The main possible options for CO_2 underground storage are illustrated in Figure 8.1. Captured CO_2 can be stored in depleted oil and gas fields, deep saline aquifers and unexploited coal seams.

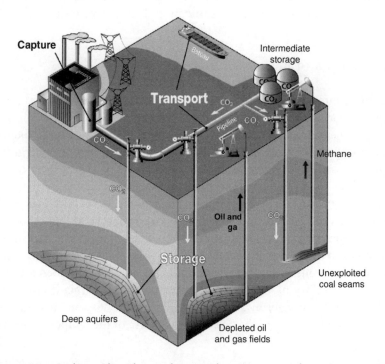

Figure 8.1 Carbon dioxide underground storage modes (*Source*: IFP–ADEME–BRGM [90]). Reproduced with permission from *Energie & Climat: Réussir la transition énergétique* by Alexandre Rojey, Éditions Technip, Paris, 2008

- ## Storage in depleted oil and gas fields

This option is particularly interesting, as it involves technologies which are already well known and geological formations which are generally well characterised.

Carbon dioxide injection in oil fields is used also for enhancing oil recovery. If CO$_2$ is injected before the field is completely depleted, in particular with the aim of EOR, it is clearly essential to separate CO$_2$ present in the production stream in order to recycle it into the reservoir.

EOR already consumes 32 million tons of CO$_2$ in the USA.

It provides an added value to the CO$_2$ which is injected and it is therefore probably the option which will be developed most rapidly in the near future.

As a further advantage, CO$_2$ is stored in geological formations which have trapped hydrocarbons over millions of years: natural confinement offered by such structures is certainly one of the most attractive aspects of this option.

- ## Storage in deep saline aquifers

Storage in deep saline aquifers has the greatest potential in terms of storage capacity. Such aquifers are distributed in many regions of the world. They are located at depths between 1000 m and 3000 m. Due to the depth of these formations and their high salt content, they cannot be used as sources of drinking or irrigation water.

These aquifers can be open or closed. The configuration of closed aquifers is identical to that of oil and gas reservoirs, which ensures effective confinement vertically and laterally.

This is the type of aquifer used to store natural gas. It certainly provides a safe option in terms of confinement, but capacity is limited.

Open aquifers lie on a horizontal or slightly inclined plane. Since they do not ensure lateral confinement this would enable CO$_2$ to migrate. This being said, their large size and low flow velocity can ensure a satisfactory confinement, assuming sufficient CO$_2$ is present. In this case, the main trapping mechanism is the dissolution of gas in water.

The increase in density which results from this dissolution tends to carry CO$_2$ towards the bottom of the aquifer. In the longer term, CO$_2$ reacts with the surrounding rock and as a result of this mineralisation process the safety of CO$_2$ trapping increases. The weak point of this solution is that little is known about the subject. This type of formation does not contain resources of interest, so it has hardly been studied. A major characterisation effort is required in order to qualify this type of aquifer for geological storage.

- ## Storage in unexploited coal seams

Storage in unexploited coal seams is based upon the capacity of coal to adsorb CO_2 preferentially instead of the methane initially present. This trapping mechanism also enables the recovery of methane, which can be recovered in producing wells, and is a potentially attractive economic benefit. The main difficulty associated with this storage solution is the low permeability of this type of formation. As a result, it is not possible to inject large quantities of CO_2 without multiplying the number of injecting wells. Such a solution is limited to unexploited coal seams in order to avoid any leakage of CO_2 through mining galleries.

Table 8.1 shows the potential CO_2 storage capacity for different geological structures (IEA data). Although current estimates are hardly precise, it appears that the potential storage capacity might be adequate for dealing with the overall world CO_2 emissions.

Table 8.1 Potential storage capacity for different storage modes

Storage options	Total capacity	
	Gt CO_2	Share of aggregate emissions in 2050 (%)
Depleted oil and gas reservoirs	920	45
Deep aquifers	400–10 000	20–500
Unexploited coal seams	40	2

A more precise evaluation of storage capacities worldwide is needed in order to assess better the overall potential CO_2 storage capacity.

Industrial applications of CO_2 capture and storage

Carbon dioxide capture and storage (CCS) can be applied to all industrial installations emitting large amounts of CO_2.

The main application expected in the future is the recovery and storage of CO$_2$ emitted by fossil fuel power plants, which amounts to 40% of CO$_2$ emitted worldwide.

Coal-fired power plants clearly represent the main target to be considered. The first way to reduce their emissions is to improve the efficiency of the plant, by increasing the pressure at which steam is generated and the temperature at which it is superheated. It requires the use of highly performing steels.

The average efficiency of coal-fired power plants is around 30 % worldwide. It can approach 45 % for the most modern installations nowadays. In the near future, it might become close to 50 % if ultra-supercritical cycles are used[2].

Emissions per kWh produced are inversely proportional to the efficiency and increasing the efficiency is the first action to undertake, before capturing and storing CO$_2$.

Nevertheless, deploying CCS is the only way to curb sharply the emissions from fossil fuel power plants in the future. In order to achieve such a goal, it is necessary to overcome obstacles which are not only technical but also economical. In the case of coal-fired power plants, CO$_2$ emissions amount to around 600 kg per MWh of electricity produced for 50 % of efficiency and 800 kg for 40 %. This has to be compared with 300 kg in the case of a gas-fired combined cycle with efficiency close to 60 %.

For a cost of € 50 for each ton of CO$_2$ not emitted to the atmosphere as a result of the application of CCS, an additional cost of € 40 per MWh has to be taken into account, which means that the cost for each MWh of electricity produced is almost doubled.

It is expected that CCS might become operational for fossil fuel power plants by 2020.

During a first phase, CCS might be used mainly in conjunction with EOR, the CO$_2$ source being provided by natural gas processing units.

During a second phase, beyond 2020, new industrial projects might involve CO$_2$ recovery from flue gases and CO$_2$ injection in deep aquifers.

CCS can also be applied in other industrial sectors which are large emitters of CO$_2$. For its production, 1 ton of steel generates 1.8 ton of CO$_2$ and 1 ton of cement generates 0.89 ton of CO$_2$.

---

[2] These are cycles in which steam reaches temperatures around 700 °C, which is much higher than the water critical temperature (374 °C), requiring the use of compatible materials.

In the future, the production of synthetic fuels might also become a major source of CO_2. For all these applications, CCS is to be considered as the main tool in the future for curbing CO_2 emissions.

In the longer term, CCS might avoid the equivalent of 6–7 Gt of CO_2 emitted each year.

Thus IEA estimates that around 6.4 Gt/year of CO_2 emissions will be avoided with the help of CCS by 2050, including 3.8 Gt/year in the sector of electricity producers, 1 Gt/year in the area of synthetic fuels production and 1.6 Gt/year in other industrial sectors [18].

In order to be able to apply industrially CCS in the future, it has to become part of the carbon emissions trading scheme, especially in the European Union which has set up the first large trading scheme. This is not yet the case and this issue is presently debated. It will require a rigorous assessment method of CO_2 emissions which can be avoided in this way.

Maintaining the competitiveness with industries abroad not submitted to the same constraints is also an important issue to be settled, especially for exporting industries, when implementing CCS. Due to its comparatively high cost, a mandatory application of CCS might lead to a relocation of activities, such as the steel industry, to countries which would not impose similar constraints. Such a result would of course be quite the opposite of the initial goal. It is therefore necessary to introduce at international level mechanisms which will avoid such a breach of the rules enabling fair competition.

Geological storage operations worldwide

Different large scale CO_2 geological storage operations have already been undertaken or are under way.

The first operation was launched in 1996 at the Sleipner site, in the North Sea, where, since that date, the Norwegian company Statoil injects 1 million tons each year of CO_2 separated from natural gas in a saline aquifer located at a depth of 1000 m below the sea floor. In 2001, a pilot project combining CO_2 storage with EOR was initiated at the Weyburn oil field in the Canadian province of Saskatchewan. The CO_2 injected comes from a coal gasification plant in North Dakota (USA) from where 5000 tons of CO_2 flow daily through a 330 km long, cross-border pipeline.

Injection of CO_2 in coal seams has also been tested. A demonstration pilot plant has been tested in Poland, within the framework of the Recopol project. An industrial CO_2 storage project has been operated since 2004 by BP, Sonatrach and Statoil at In Salah, in Algeria, where 1 million tons of CO_2 per year are being injected into a deeper part of the same geological layer that contains the gas which is produced.

Numerous EOR projects involving CO$_2$ injection are being investigated worldwide. Many such projects operate or are launched in the USA where 70 such projects exist already. New projects are also being investigated in Europe. In the north of Scotland, BP has been investigating the injection of 1.3 million tons of CO$_2$ per year produced by a 350 MW power plant located at Peterhead in northeast Scotland.

The project involved the production of a mixture of hydrogen and CO$_2$ from natural gas, the hydrogen thus produced being used in a combined cycle after separation of CO$_2$ but, presently, it is delayed or it may perhaps even be cancelled.

Shell and Statoil are investigating an EOR project involving the injection in the Norwegian field of Draugen of 2.5 million tons of CO$_2$ recovered from a gas-fired 860 MW power plant.

In Germany, electricity producers have announced their intention to build pilot plants and demonstration units for validating the application of CCS to coal-fired power plants. The Vattenfall Company is building a 30 MW pilot plant for testing oxy-fuel combustion in the 'Schwarze Pumpe' industrial zone on the Brandenburg/Saxony border. Other electricity producers such as RWE and E.ON have also announced their intention to build demonstration units.

Research and development programmes are needed to develop safer and more cost-effective CCS installations.

The main obstacles arise from economics: CCS costs are presently around 50 € per ton of avoided CO$_2$. Further cost reductions are required in order to ensure CCS deployment.

It is also essential to demonstrate the safety and reliability of geological storage over very long periods of time (centuries or even thousands of years). For that purpose, it is necessary to improve the knowledge of CO$_2$ behaviour in geological reservoirs.

Legal and social framework for geological storage

Carbon dioxide geological storage currently suffers from a lack of a regulatory framework, especially in the case of onshore operations.

Appropriate legislation applying to long-term storage has yet to be defined. This situation slows down the deployment of CCS and has to be clarified.

Onshore, existing regulations which can be applied to geological storage vary from country to country and generally are not defined for CO$_2$ geological storage as such.

The only case which does not require a new regulatory framework is CO$_2$ injection in hydrocarbon reservoirs for EOR.

In all other cases, existing legislation needs to be adapted unless new specific rules are adopted, in order to take into account CO_2 geological storage.

Offshore, underground storage is regulated by the UNCLOS (United Nations Convention on the Law of the Sea) Convention, which defines general rules, the London Convention (Convention on the Prevention of Marine Pollution by Dumping of Wastes and Other Matter, signed in 1972), which deals with the protection of the marine environment and the OSPAR Convention (Convention for the Protection of the Marine Environment of the Northeast Atlantic, 1992).

The purpose of these treaties was to protect the marine ecosystem from possible pollution and they were not specifically taking into account the storage of CO_2.

In 2006, an amendment of the London Convention now allows CO_2 injection in underground reservoirs offshore subject to certain conditions concerning implementation [93].

At the European level, various actions are under way in order to prepare the future directives which will come into effect to authorise geological storage projects.

Carbon dioxide geological storage is not yet taken into account in the carbon trading mechanisms, whereas it seems a prerequisite for its future development.

This issue is actively discussed within the different competent institutions which are involved, especially within the United Nations (United Nations Framework Convention on Climate Change or UNFCC) and within the European Commission.

Beyond any regulation that might be put in place, a large scale deployment of CO_2 underground storage requires public acceptance from people living nearby. This implies the need to carry out investigations to ensure the complete safety and reliability of underground storage. Good information and dialogue with local communities are also needed.

Longer term perspectives: mineral sequestration and CO_2 recycling

Mineral sequestration of CO_2 consists of transforming it into a stable substance through a carbonation reaction with a rock presenting a basic activity.

The main difficulty stems from the fact that such reactions are comparatively slow. Two channels are currently being explored:

- *Ex situ* mineral sequestration is operated above ground in an industrial installation by reacting CO_2 with ground rocks or with solid waste. The main drawback is the need to manipulate, grind and store considerable volumes of solid materials. Different solid phases have been investigated for use in such a process.

   Typically, CO_2 is reacted with crushed olivine and serpentine (magnesium-rich silicate rocks) from a mine. Industrial wastes, such as blast furnace slag, composed of iron and calcium silicates, can also be used.

- *In situ* mineral sequestration is operated by injecting CO_2 in a natural setting where basic rocks of magmatic origin, such as basalt, are present. Such an operation is performed in a very similar way to that used in the case of geological storage. The carbonation reaction between CO_2 and the surrounding rock occurs very slowly and results in the formation of a stable compound. The porosity of magmatic rocks such as basalts is generally low and it is therefore difficult to inject large amounts of CO_2.

   The porosity and injectivity of such rocks can be increased through fracturation operations, but such operations are costly and it appears difficult to use them on a large scale in the near future.

Carbon dioxide recycling includes all possible uses in industrial or biological processes. Unfortunately the quantities of CO_2 which can be recycled in that way are limited. Currently, the main application in the chemical industry is the production of urea, which requires around 80 million tons of CO_2 per year [92]. A significant outlet is provided by the agro-food industry which, in Europe, consumes around 2.7 million tons of CO_2. A CO_2 rich atmosphere is used for accelerating the growth of plants cultivated in greenhouses. This outlet is also limited, but it is possible to widen the applications of the biomass produced in a CO_2 enriched atmosphere by investigating new applications in the agro-food, chemical and energy industries.

The production of algae in bioreactors is presently considered for sequestering CO_2 produced by a fossil fuel power plant. Such algae can then be converted into biofuels (diesel from the fatty fractions and/or ethanol from sugars derived from the biomass). Despite the fact that this area is very active, it remains difficult to ascertain the economic competitiveness of these developments.

It is necessary also to assess the overall carbon balance: the combustion of such biofuels cannot be considered as carbon neutral, because the carbon used for growing the biomass is provided by a fossil fuel and not recovered from the atmosphere.

EOR is currently the main large scale application of CO_2 injection and it should continue to grow in the future.

The role of CO_2 geological storage

CCS should contribute significantly in the future to the reduction of CO_2 emissions.

It is typically a 'transition technology' to be used for providing an answer to the risk of climate change for the period during which fossil fuels will still contribute significantly to the energy supply.

Geological storage should be considered as a prerequisite before any new large scale development of the use of coal for electricity generation. Without CCS, the comeback of coal might result in even higher CO_2 emissions than those which can be predicted from an extrapolation of the present trend.

As large-scale industrial deployment of CCS for fossil-fuel power plants is considered only beyond 2020, the present development of electricity generation from coal-fired power plants represents a real risk.

It is therefore necessary to accelerate the deployment of these options. This requires the intensification of research and development activities aimed at reducing costs and ensuring the long term safety of geological storage.

It is also essential to set up as soon as possible the regulatory framework needed for implementing CO_2 geological storage and to take CCS into account within the carbon trading mechanisms.

Research actions required

... **For capturing and storing CO$_2$**

- **New capture processes**
  - More cost-effective processes (reduction of the investment and of the energy consumption)
  - Direct oxygen transfer to the combustion stage ('chemical looping', adsorption–desorption, membranes)
- **Improving the reliability of underground storage**
  - Development of predictive modelling tools
  - Monitoring methods (seismic and geochemical)
- **Investigation of new underground storage sites**
  - Identifying new underground storage sites (especially deep aquifers)
  - Alternative storage modes (including *in situ* mineralisation)
  - Recycling captured carbon
  - Use of captured carbon for producing materials, chemicals or synthetic fuels (consuming a low carbon energy)
- **Carbon sinks**
  - Rapidly growing biomass for improving the kinetics of CO$_2$ capture
  - Identifying new sites for carbon sinks

9

How to Ensure the Energy Transition?

The conditions required for a successful transition

Can the different options presented previously be used for ensuring a satisfactory energy transition?

The conditions required for such a successful transition have already been discussed. The limitation of CO_2 emissions is certainly the main constraint. Despite all the uncertainties which remain, reasonable assumptions have to be made for the future. The objective adopted by the European Union to limit to 2 °C the increase of the mean temperature appears as the most appropriate. As already discussed, CO_2 emissions must stay at a level compatible with a CO_2 concentration in the atmosphere of 450 ppm by 2050. In order not to exceed that level, it is necessary by 2050 to divide world emissions of CO_2 by a factor of around 2, as compared with the present level.

At the same time, we might be faced with a limitation of the world production of hydrocarbons. When dealing with such a situation, it is necessary to avoid both a major economic crisis and an environmental catastrophe. Although challenging, the task is not impossible, by initiating appropriate actions without delay. Whilst the extrapolation of the present trend leads to an intolerable situation, it is still possible, by deploying the various means described previously, to implement an evolution scenario reaching an acceptable situation by 2050.

Energy and Climate: How to achieve a successful energy transition Alexandre Rojey
Copyright © 2009 Society of Chemical Industry

In order to succeed, it will be necessary, besides technological progress, to introduce creativity and innovation in all areas where energy is used, transforming progressively the way of life itself.

It is therefore the project of a new society that we have to consider.

Implementing an acceptable evolution scenario

By using the four action points presented in the previous chapters (reducing the consumption of energy, reducing its carbon content, managing fossil fuel supplies, and capturing and storing emitted CO_2), it is possible to satisfy world energy needs, while keeping CO_2 emissions within acceptable limits.

In order to demonstrate that it is feasible, let us examine evolution scenarios from now to 2050. The year 2050 represents a limit for any present extrapolation and is also an important landmark for the global CO_2 balance. It is at this date at the latest that we need to stabilise the concentration of CO_2 in the atmosphere (which requires, as it has already been mentioned, an action plan to be started now).

Starting from the present level of consumption (considering 2006 as the reference year), the BAU ('Business as usual') scenario, is obtained by extrapolating the present trends. According to this scenario, world energy consumption will double within less than fifty years, from 11.7 Gtoe/year in 2006 to 22.7 Gtoe/year in 2050. The major share (80%) is still supplied by fossil fuels; coal represents 30% of the primary energy supply, oil 32 % and natural gas 22 %. The corresponding CO_2 emissions rise from 27.9 Gt/year in 2006 to 52 Gt/year in 2050. This scenario is clearly intolerable. However, it is not the most pessimistic one.

The second alternative (Alt) scenario aims to fulfil the same needs, but with an emission level reduced to 14.5 Gt/year in 2050, which is a reduction factor of 3.6 when compared with the reference scenario (BAU) (Figure 9.1).

The Alt scenario is obtained by acting upon the four points described previously:

- Improving the energy efficiency results by 2050 in reducing by 44% the annual energy consumption (when compared with the BAU scenario), from a consumption of 22.7 Gtoe/year to a consumption of 12.6 Gtoe/year.
- The total share of fossil fuels decreases from 80 to 59%.

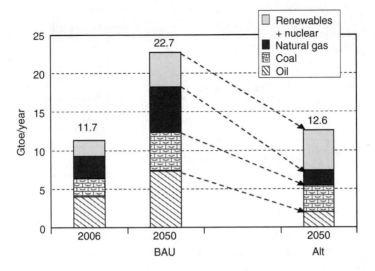

Figure 9.1 Comparison between the reference (BAU) scenario and the alternative (Alt) scenario

- Coal consumption is divided by three as compared with the BAU scenario and represents only 16% of the total energy consumption in the Alt scenario.
- The consumption of hydrocarbons, although strongly reduced when compared with the BAU scenario, remains still close to the present level, with a significant increase in the share of natural gas, which implies a significant effort to maintain the corresponding level of production.

The Alt scenario is consistent with the assumption of oil consumption reaching a ceiling before 2050, declining later, and coming back to a level lower than the present one by 2050. It implies the use of oil mainly for engine fuels production to achieve a large reduction in demand and the need to develop alternatives (biofuels, plug-in hybrids and possibly hydrogen as an energy vector by 2050).

The consumption of natural gas is supposed to increase until 2030 and then to become stable according to a consumption plateau.

Reaching the required goals

The growth according to these two scenarios is represented by the graph in Figure 9.2. In this graph, the CO_2 emissions expressed in billions of

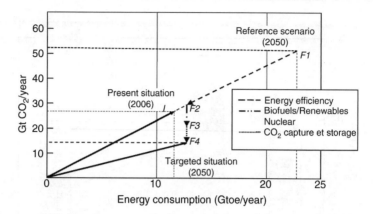

Figure 9.2 Evolution scenario of the world energy demand and of CO_2 emissions between now and 2050. Reproduced with permission from *Energie & Climat: Réussir la transition énergétique* by Alexandre Rojey, Éditions Technip, Paris, 2008

tons per year are plotted versus the energy consumption expressed in billions of tons of oil equivalent per year.

The reduction of CO_2 emissions is achieved by using three of the four action points already mentioned:

- Improving energy efficiency leads by 2050 to a reduction of around 23 Gt/year of CO_2 emissions (moving from point F1 to point F2).
- Increasing the share of low carbon energy in the energy mix makes a reduction of CO_2 emissions of around 7.5 Gt/year possible (moving from point F2 to point F3).
- CCS and carbon sinks lead to a further reduction of CO_2 emissions of around 7 Gt/year (moving from point F3 to point F4). Carbon sinks (voluntary reforestation operations or biomass production for the purpose of removing carbon from the atmosphere) represent only a limited share of this total amount, not exceeding around 1 Gt/year. The deployment of CCS will require a huge effort. Indeed, in the case of an operation like the one at Sleipner, the flow rate of CO_2 injected underground is 1 Mt/year and on a worldwide scale the equivalent of 6000 to 7000 operations of that type are needed by 2050 to reach the required reduction in level of CO_2 emissions.

Figure 9.3 presents the evolution with time of worldwide CO_2 emissions in the reference (BAU) scenario and in the Alt scenario. It shows the relative impact of the different factors involved.

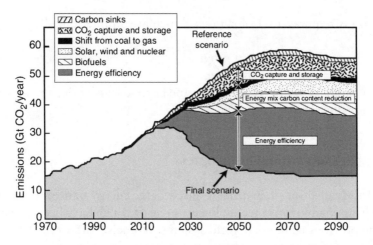

Figure 9.3 Evolution scenario of worldwide CO_2 emissions

Figure 9.3 is based upon a scenario established at a European level, presented in a report of the European Environment Agency, based upon assumptions close to those presented previously [95, 96].

Carbon dioxide emissions reach a maximum by 2020, and then decrease at a level close to half the present level by 2050.

Future outlook

The Alt scenario shows that it is possible to succeed, but it cannot be implemented without a strong political will and wide support from public opinion. Therefore the question arises whether it is realistic and what are its real chances of success.

The main difficulty is to conclude a successful agreement on a worldwide level for implementing an international action plan.

Public opinion will be most influenced by catastrophic events, linked to climate change or to a sudden disruption of energy supplies. A true understanding of the potential risks might occur too late, which would not leave enough time for anticipating any necessary action properly.

The urgency of the actions to be undertaken is clear however. The sooner we are able to start them, the more likely we will be to avoid catastrophic consequences resulting from inaction. It is necessary therefore to move forward, bearing in mind that any progress made will have a significant impact in the future.

Success requires a set of actions, which are all interlinked:

- ensuring technological progress;
- introducing a new lifestyle;
- defining new governance rules and beyond legislation, a new project for our society.

Ensuring technological progress

As already mentioned, in order to succeed, the energy transition will have to rely upon innovative technologies, some of which have been described in the previous chapters (hybrid systems, biofuels, batteries, CCS, etc.).

These different technologies will be introduced progressively and will need to be deployed through successive stages, in order to ensure a transition without any interruption.

During a first phase, improving energy efficiency is the main solution for reducing CO_2 emissions before 2030 and to respond to the limitation of the oil and gas production. It will be a key factor for the next two decades.

Then, between 2015 and 2030, subject to a rapid and massive move in the right direction, other options such as second-generation biofuels, electricity generation from low carbon sources and CCS will play an increasing role.

In the longer term, by 2030 to 2050, more radical technological changes are expected: namely fourth generation nuclear power plants, large scale photovoltaic electricity generation, distributed energy storage systems and use of hydrogen as an energy vector.

Table 9.1 presents the different phases of implementation of the energy transition technologies, according to what seems likely now. It is of course possible to be more optimistic or pessimistic about this forecast.

Table 9.1   Future implementation of the energy transition technologies

Short term (before 2020)	Mid term (2015–2030)	Long term (2030–2050)
Increased energy efficiency	Hybrid systems	Hydrogen as an energy vector
New materials	Second-generation biofuels	Fourth generation nuclear power plants
Control and regulation systems	CCS	Electricity storage
Energy from biomass		Photovoltaic systems

The development of clean and performing technologies for the production and use of energy will be favoured by the convergence of technological progress in the following areas:

- *Advanced materials and nanotechnologies* New materials will be used for reducing friction losses, reaching higher temperatures and efficiencies, improving energy storage or developing new photovoltaic systems.

    New catalysts and separation membranes will be used for developing industrial processes that are more energy-efficient and more environmentally friendly.
- *Information technologies and sensors* New developments in the area of information technologies and sensors will help to optimise on-line propulsion and energy conversion systems.

    Such technologies are already widely used by the car manufacturing industry: on board electronics control the engine running and can minimise the fuel consumption. On-line optimisation systems can also be used to minimise energy consumption in industry and housing.
- *Biotechnologies* Progress in biochemistry will help to produce biofuels and more generally energy from biomass. This will lead also to the development of 'green' chemistry, to ensure better management of drinking and irrigation water and to recycle waste.

Such transition technologies represent a major economic stake. These new developments appear more and more as future sources for profit. Clean technologies generate a new wave of 'cleantech' investments throughout the world and are considered to be a major new opportunity [94, 98]. Currently in the USA, and more especially in California, there is such a rapid and sudden surge in risk-capital that it creates the risk of a financial 'bubble' similar to that created by the internet [99].

Even if some excessive speculative move has to be observed with care, this situation looks promising as it demonstrates the interest of investors and should help to develop innovative technologies.

Working out the energy transition requires a pragmatic step-by-step approach. Each step has to be planned in order to move towards the desired objectives and to prepare the following steps.

It is difficult to forecast now the technological progress which might occur by 2050 and beyond.

The uncertainties arise from the difficulty of predicting both the rapidity of the incremental progress which is expected and the occurrence of breakthrough innovations, which would radically change preconceived ideas. In any case, transition technologies should help to prepare for the future.

In the area of energy production, a diversification of the energy sources to be used is expected. During the transition period, electricity generation from fossil fuels, integrating CCS, will coexist with generation from nuclear and renewable energies, which will represent a growing share.

Progress expected in the area of photovoltaic electricity generation should lead to an increase in the share of decentralised electricity production. The cost of photovoltaic electricity generation is not sensitive to scaling up, so the transportation cost becomes comparatively more important, which favours decentralised systems. An increasing use of intermittent energy sources, such as wind and solar energies, will require in the longer term the deployment of electricity storage means. Energy storage is therefore a key technology for the future, as its deployment will be essential for increasing substantially the share of renewable energy sources.

It is also necessary to reconsider the whole system of transport for people and for goods.

Thus, in the future, vehicles adapted to a certain environment might become more widespread and, for instance, the centre of urban areas might be restricted to some nonpolluting vehicles only. The advantages of plug-in hybrid vehicles as a future solution have already been mentioned. The sharp consumption reduction, which would result from the use of such an option, might also facilitate the use of biofuels during the transition period. The biomass resources required for supplying a significant share of the engine fuels needed would be thus significantly reduced. This technology should help also the development of electric vehicles, if significant progress is achieved in the area of batteries to ensure adequate mileage.

Obstacles to overcome for using hydrogen as a fuel have already been mentioned, but should not stop research and development activities as new breakthrough innovations might occur in the future and help to achieve the required objectives.

Air transport raises difficult questions. Reducing fuel consumption through the introduction of new materials and the improvement of propulsion technologies is an important factor for future progress, but the gains which can be achieved will be most certainly more than compensated by the increase in traffic. In the future, kerosene might be at least partly produced from biomass. For economic reasons and taking into account the limited availability of biomass, such a production will probably supply only a fraction of the needs.

Hydrogen looks an attractive option for air transport. The high energy density achieved with hydrogen with respect to on board weight is a real asset, in addition to its environmental qualities. Still, its use will be difficult to implement, as it requires a completely new design of the aircraft. Besides the difficulties linked to the production and distribution of liquid hydrogen, the aircraft itself will have to accommodate much bigger reservoirs. Nevertheless, if safety can be ensured in a satisfactory way, this option has to be seriously considered as in the longer term, there is no other serious alternative, apart from a stringent limitation on air travel and transport, which cannot be completely discarded.

It is necessary to facilitate the progressive introduction of future solutions which can be anticipated now. Hybrid and flexible technologies are well adapted for the transition period, as they can help to operate the changes which will be necessary during this period.

A new lifestyle

As already stressed, the present situation requires a thorough transformation of our economic model. In order to perform such a transformation under acceptable conditions, innovating technologies are needed. New pathways have to be explored, opening a wide field of investigation for research.

Technological progress alone is not enough. The changes brought by technology have to be used within a more general framework, involving an evolution in the way of life and mentalities.

Such a project requires the right policy and governance on a worldwide scale.

It is only by inventing new solutions for improving the quality of life while reducing energy consumption, that it will be possible to get the support of public opinion. Any pollution, chemical, acoustic or even visual threatens the quality of life. Looking for a harmonious and healthy setting is a concern of growing importance for everybody. Well developed countries have to show how to improve the quality of the environment and implement sustainable solutions. Emerging countries will thus be encouraged to adopt such a new paradigm, rather than undergo all the development stages that have been followed by the richer countries.

What happens now will be decisive for the years 2030–2050. A strong political will is essential for adopting the appropriate legislation and framework required to implement the changes which are needed.

A new project for the future society

The amplitude of change in lifestyle, combined with the introduction of innovative technologies, will be such that it has to be considered as a new project for the future society.

It has been proposed to call this new society the 'Factor 4 Society' [51].

New concepts for housing and transport will result not only in a reduction in energy consumption but also in a better quality of life. The introduction of more flexible transport means with the help of microelectronics and more intelligent software will make their use more attractive and help to reduce the use of individual cars. The desire to live in a more pleasant setting will help to develop pedestrian districts. Liberating town centres from pollution and noise resulting from car traffic will help to create more concentrated housing, consuming less space and energy.

Further progress in telecommunications will help to limit the need for physical moves. Looking for a more healthy physical condition will be further motivation for discarding excessive use of the individual car and will encourage the use of walking or cycling. It will favour also the circulation of biological products, of better quality, obtained by using less intensive agriculture.

In a society in which knowledge becomes more and more important, cultural goods and services will become increasingly important. The value of information is not linked to the quantity of resources it requires. A single DVD disc can hold the equivalent of a whole library. The internet opens the way to an almost unlimited universe of artistic and intellectual creation.

Once our basic needs are satisfied, virtual worlds will be able to fulfil a large part of our requirements for intellectual discovery and entertainment, without any significant consumption of energy or resources.

The creation of value does not necessarily imply the consumption of supplementary resources. Artistic values, the beauty of landscapes or human creation can be appreciated at their real cost, even if it is difficult to quantify.

Conclusion

A number of conclusions stand out clearly from the analysis presented in this book.

First, since the current energy system, which results in massive CO_2 emissions and high dependency on oil supplies, is not sustainable, it is essential to initiate a transition.

The measures to be taken are urgent, while substitution of fossil energies by alternative, non-CO_2-emitting energies will take time.

It is absolutely vital to rapidly implement a set of means specifically adapted to the transition period ahead.

Four action points have been identified:

- Reduction of energy consumption is the first type of action to implement. This will relieve the tensions over hydrocarbon supplies and reduce CO_2 emissions. In addition, significant results can be obtained by applying existing solutions.
- Energy decarbonation, although a critical objective for the future, will be difficult to complete in the short term. Transition solutions (diversification of the energy *mix*, biofuels, etc.) must be set up. The transition period will be marked by the development of *hybrid solutions* combining different forms of energy.
- Fossil energy supplies will continue to play a major role during the transition period. They will allow the short respite needed to deploy the alternative energy sources. We must diversify the supply sources and, above all, avoid resorting to the fuels most detrimental to the environment.

Energy and Climate: How to achieve a successful energy transition   Alexandre Rojey
Copyright © 2009 Society of Chemical Industry

- Carbon dioxide capture and storage must be implemented as quickly as possible; in particular, this is a prerequisite to the more widespread use of coal.

Innovation has a key role to play to achieve the transformations required. New solutions must be invented or developed. These mutations represent a major opportunity for those with sufficient foresight to anticipate them.

Technical solutions alone will not be sufficient, however. They can only develop efficiently in the context of suitable state policies conducted with determination.

Since the problems facing us affect the entire planet, these policies must be implemented through international agreement, involving as many countries as possible. Nonetheless, the need for greater international coordination must not give rise to a 'wait-and-see' attitude at national level: the initiatives taken today towards new solutions can only help the decisions and choices to be made at world level.

Lastly, it is essential to adapt our ways of life. We must also call upon imagination and innovation to invent new modes of habitat and mobility. More than an evolution, we are talking about a revolution. We have an obligation to succeed, for the planet and for our children. It demands massive and unfailing participation from all citizens, in other words, by every one of us.

Appendices

Appendix 1 – Energy and power

- Irrespective of the initial source, the first law of thermodynamics states that energy is conserved and converted into equivalent quantities of work or heat.

 The second law of thermodynamics states that when heat energy is converted into work, some is always lost to the surroundings as heat and cannot be reused. The 'quality' of the thermal energy depends on the temperature. The lower the temperature and the more it approaches ambient, the lower the quality.

 The legal, scientific unit (Système International or SI) used for the measurement of energy is the joule (J).

 In an economic context, the tonne of oil equivalent (toe) is frequently used, and sometimes the tonne of coal equivalent (tce).

 The kWh is another common unit, corresponding to the amount of work or energy generated when one kilowatt of power is supplied for a time of one hour.

- Power is the quantity of energy per unit of time or the rate of doing work, which is the same thing. The legal, scientific unit of power (SI unit) is the watt (W), which is one joule per second. Multiples of the watt are also used: kW (1000 W), MW (1000 kW), GW (1000 MW) and TW (1000 GW).

 A watt peak (Wp) designates the maximum power delivered by a device powered by an intermittent energy source (e.g. a photocell).

Appendix 2 – Conversions and equivalences between units

1 W	$= 1$ J/s
1 kWh	$= 3.6$ MJ
1 MWh $= 1000$ kWh	$= 0.086$ toe
1 toe (tonne oil equivalent)	$= 41.8$ GJ $= 7.33$ b
1 tonne of coal (anthracite)	$= 0.67$ toe
1 tonne of coal (lignite)	$= 0.33$ toe
1000 m^3 of natural gas	$= 0.9$ toe
1 tonne of LNG	$= 1.23$ toe
1 barrel (b)	$= 159$ litre
1 barrel per day (b/d)	$= 49.8$ t/year

Appendix 3 – Abbreviations and acronyms

- **Multiples of 1000**
  k   kilo (10^3)
  M   mega (10^6)
  G   giga (10^9)
  T   tera (10^{12})
- **Other abbreviations and acronyms**

ADEME	Agence De l'Environnement et de la Maîtrise de l'Énergie (*French Environment and Energy Management Agency*)
AFC	Alkaline Fuel Cell
ASPO	Association for the Study of Peak Oil & Gas
ATIC	Association Technique de l'Importation Charbonnière (*Technical Association for Coal Imports*)
BAU	Business As Usual
BRGM	Bureau de Recherches Géologiques et Minières (*Geological and Mining Research Bureau*)
BTL	Biomass To Liquids
CCS	CO_2 Capture and Storage
CDM	Clean Development Mechanism
CEA	Commissariat à l'Énergie Atomique (*Atomic Energy Commission*)
CFC	ChloroFluoroCarbon
CIF	Cost, Insurance and Freight

CNG	Compressed Natural Gas
CTL	Coal To Liquids
DC	Developing Countries
DEFC	Direct Ethanol Fuel Cell
DMFC	Direct Methanol Fuel Cell
EPR	European Pressurised Reactor
ETBE	Ethyl Tert-Butyl Ether
ETS	Emissions Trading Scheme
EU	European Union (EU-15, with the first fifteen countries; EU-25, with twenty-five countries)
FACTS	Flexible Alternative Transmission Systems
FNR	Fast Neutron Reactor
GDP	Gross Domestic Product
GHG	Green House Gas
GMO	Genetically Modified Organism
GNP	Gross National Product
GPI	Genuine Progress Indicator
GTL	Gas To Liquids
HDI	Human Development Index
IEA	International Energy Agency
IPCC	Intergovernmental Panel on Climate Change
ITER	International Thermonuclear Experimental Reactor
JI	Joint Implementation
LED	Light Emitting Diode
LNG	Liquefied Natural Gas
LPG	Liquefied Petroleum Gas
MEA	MonoEthanolAmine
MIT	Massachusetts Institute of Technology
MOFC	Molten Carbonate Fuel Cell
Mox	Mixed oxide (nuclear fuel containing recycled plutonium)
NO_x	Nitrogen Oxides
NRE	New/Renewable Energies
OPEC	Organisation of Petroleum Exporting Countries
OSPAR	Designation of the Oslo–Paris Convention guiding international cooperation on the protection of the marine environment of the northeast Atlantic
PAH	Polycyclic Aromatic Hydrocarbon
PAN	Peroxyacetylnitrate
PEMFC	Proton Exchange Membrane Fuel Cell
PHV	Plug-in Hybrid Vehicle

PM_x	Particles of diameter less than x μm
ppm	parts per million
PWR	Pressurised Water Reactor
SOFC	Solid Oxide Fuel Cell
SO_x	Sulphur oxides
UNCLOS	United Nations Convention on the Law Of the Sea
UNEP	United Nations Environment Programme
UNFCC	United Nations Framework convention on Climate Change
UNO	United Nations Organisation
VNG	Vehicle Natural Gas
VOC	Volatile Organic Compound
VOME	Vegetable Oil Methyl Ester
WMO	World Meteorological Organisation

References

[1] *World Energy Outlook*, IEA, 2008.

[2] *Key World Energy Statistics*, IEA, 2007.

[3] *Energy to 2050*, IEA, 2003.

[4] Jean-Pierre Favennec, *Géopolitique de l'énergie*, Éditions Technip, 2007.

[5] Jean-Marie Chevalier, *Les grandes batailles de l'énergie*, Folio Actuel, 2004.

[6] Stéphanie Monjon, *Mondialisation et environnement*, Développement et Environnement, Cahiers français 337, March–April 2007.

[7] Al Gore, *Earth in the Balance*, Earth Scan Limited, Revised Edition, 2007.

[8] *BP Statistical Review*, 2008.

[9] Dominique Dron, Les enjeux d'un climat soutenable, in *Regards sur la Terre*, Pierre Jacquet and Laurence Tubiana, Presses de Sciences Po, Paris, 2006.

[10] Jean-Marc Jancovici, *L'avenir climatique*, Seuil, 2002.

[11] Tim Flannery, *The Weathermakers*, Grove Press, 2005.

[12] Adolphe Nicolas, *Futur empoisonné*, Belin, Pour la Science, 2007.

[13] Nicolas Stern, *Review on the Economics of Climate Change*, 2006.

[14] Yves Cochet, *Pétrole apocalypse*, 2005.

[15] James H. Kunstler, *The Long Emergency*, Atlantic Books, London, 2006.

[16] Pierre Lafitte, Claude Saunier, *Changement climatique et transition énergétique: dépasser la crise*, rapport OPECST, 2006.

[17] Pierre-René Bauquis, *Oil & Gas Science and Technology*, 56, 4, 2001.

[18] *Energy Technology Perspectives*, IEA, 2006.

[19] *IPCC Special Report on Emissions Scenarios*, edited by Nebojsa Nakicenovic and Rob Swart, IPCC, COP 6, 2000–2001.

[20] Richard S. Lindzen, *Why so gloomy*, Special Report: Living with Climate Change, Newsweek, 16–23 April, 2007.

[21] George Monbiot, *Heat – How to Stop the Planet from Burning*, Anchor Canada, 2007.

Energy and Climate: How to achieve a successful energy transition   Alexandre Rojey
Copyright © 2009 Society of Chemical Industry

[22] Gary Yohe, Elizabeth Malone, Antoinette Brenkert, Michael Schlesinger, Henk Meij, Xiaoshi Xing, Daniel Lee, *A Synthetic Assessment of the Global Distribution of Vulnerability to Climate Change from the IPCC Perspective that Reflects Exposure and Adaptative Capacity*, CIESIN, Columbia University, New York, April 2006.

[23] Le Cercle des économistes, Erik Orsenna, *Un monde de ressources rares*, Perrin/Descartes and Cie, 2007.

[24] Ervin Laszlo, *The Chaos Point. The World at the Crossroads*, Piatkus, 2006.

[25] Jared Diamond, *Collapse – How Societies Choose to Fail or Succeed*, Penguin Books, 2006.

[26] Thomas Hover-Dixon, *The Upside of Down-Catastrophe, Creativity and the Renewal of Civilisation*, Souvenir Press, 2006.

[27] Dominique Bourg, Gilles-Laurent Rayssac, *Le développement durable. Maintenant ou jamais*, Découvertes Gallimard, Sciences et Techniques, 2006.

[28] Donella H. Meadows *et al.*, *The Limits to Growth*, Earth Island, 1972.

[29] Gilles Rotillon, *Les différentes approches du développement durable*, Développement et Environnement, Cahiers français 337, March–April 2007.

[30] Étude WETO, *Énergie, technologie et politique climatique: les perspectives mondiales à l'horizon 2030*.

[31] Peter W. G. Newman, Jeffrey R. Kenworthy, *Sustainability and Cities: Overcoming Automobile Dependence*, Island Press, Washington, DC, 1999.

[32] Jean Marc Jancovici, Alain Grandjean, *Le plein, s'il vous plait*, Éditions du Seuil, 2006.

[33] Jean-Charles Hourcade, Frédéric Ghersi, *La taxe carbone: une bonne idée à ne pas gâcher*, Pour la Science, no. 54, January– March 2007.

[34] Patrick Criqui, *Effets de serre: quelques scénarios*, Futuribles, January 2006.

[35] Rapport du Groupe de travail, *Division par quatre des émissions de gaz à effet de serre de la France à l'horizon 2050*, under the chairmanship of Christian de Boissieu.

[36] Jean-Pierre Dupuy, *Pour un catastrophisme éclairé*, Éditions du Seuil, 2002.

[37] *Énergie – Des choix qui engagent pour cent ans*, Le Monde Diplomatique, January 2005.

[38] *L'intensité énergétique de la croissance chinoise – Tendances et enjeux*, LEPII – EPE, Colloque international 'La Chine au cœur de la croissance mondiale: concurrence, opportunités, restructuration de réseaux économiques', CREM, Université de Rennes 1, 1–2 December 2005.

[39] Gilles Pison, *6,5 milliards d'hommes aujourd'hui, combien demain?*, Revue du Palais de la Découverte, no. 347, April 2007.

[40] *Hypergreen: une résille de béton pour une tour écologique*, Les Échos, special 'Sustainable Development' Edition, 28 March 2007.

[41] Pierre-Noël Giraud, Benoit Lefèvre, Transport et urbanisme, le défi des villes du Sud, in *Regards sur la Terre*, P. Jacquet and L. Tubiana, Presses de Sciences Po, Paris, 2006.

[42] Cédric Philibert, *Transport, énergies et facteur 4*, DATAR, Études et prospectives, December 2005.

[43] *Deux technologies pour réduire les consommations d'éclairage, aujourd'hui et dans les années 2020*, Énergie et développement durable, no. 14, November–December 2006.

[44] Patrick Piro, *Guide des énergies vertes pour la maison*, Terre vivante, 2006.

[45] Amory B. Lovins, E. Kyle Datta, Odd-Even Bustnes, Jonathan G. Koomey, Nathan J. Glasgow *Winning the Oil Endgame*, Rocky Mountain Institute, 2004.

[46] Philippe Pinchon, *Futures évolutions des motorisations dans l'automobile*, Réalités industrielles, Annales des Mines, November 2003.

[47] Pierre-René Bauquis, *Quelles énergies pour les transports au XXIe siècle*, Les cahiers de l'économie, no. 55, Série Analyses et Synthèses, ENSPM-IFP, October 2004.

[48] Michel Griffon, *Une période de transition*, Les Échos, special 'Sustainable Development' Edition, 28 March 2007.

[49] Jacques Astier, *Quelques réflexions sur la récupération et le recyclage des ferrailles en sidérurgie*, Les techniques de l'industrie minérale, December 2005.

[50] Nicolas Hulot, *Pour un pacte écologique*, Calmann-Lévy, 2006.

[51] Pierre Matarasso, *Le 'facteur 4': un projet de société*, Pour la Science, no. 54, January–March 2007.

[52] Franck Carré, Jean-Claude Petit, *Nucléaire: la Génération IV*, Futuribles, January 2006.

[53] Alain Leudet, *Pourquoi et comment concevoir des systèmes nucléaires de quatrième génération?*, Systèmes nucléaires du futur, Génération IV, Clefs CEA, no. 55, summer 2007.

[54] Georges Capus, *Que savons-nous des ressources mondiales d'uranium?*, Systèmes nucléaires du futur, Génération IV, Clefs CEA, no. 55, summer 2007.

[55] Bernard Bonnin, Frank Carré, *Comment les systèmes nucléaires de 4e génération se déploieront-ils?*, Systèmes nucléaires du futur, Génération IV, Clefs CEA, no. 55, summer 2007.

[56] Sylvain David, *Le nucléaire de fission pour le futur*, Découverte, no. 344–345, January–February 2007.

[57] Bertrand Barré, Pierre-René Bauquis, *L'énergie nucléaire*, Éditions Hirlé, 2007.

[58] Gérard Claudet, La valorisation énergétique de la biomasse in *L'énergie de demain*, EDP Sciences, 2005.

[59] Paul Gipe, *Le grand livre de l'éolien*, Observ'ER, Éditions Le Moniteur, 2007.

[60] *Étude Eurostaf*, Les Échos, 8 August 2007.

[61] Patrick Jourde, Le photovoltaïque: les filières, les marchés, les perspectives in *L'énergie de demain*, EDP Sciences, 2005.

[62] *Recherche solaire 2006*, Systèmes solaires, l'observateur des énergies renouvelables, July 2006.

[63] Daniel Lincot, *La conversion photovoltaïque de l'énergie solaire*, Découverte, no. 344–345, January–February 2007.

[64] ECRIN, *Énergies alternatives*, Jean Bonal and Pierre Rossetti, Omniscience, 2007.

[65] Felix A. Peuser, Karl-Heinz Remmers, Martin Schnauss, *Installations solaires thermiques – Conception et mise en œuvre*, Systèmes solaires, Le Moniteur, 2005.

[66] *Le baromètre biomasse solide*, Systèmes solaires, l'observateur des énergies renouvelables, November–December 2006.

[67] Daniel Ballerini, *Les biocarburants: état des lieux, perspectives et enjeux du développement*, Éditions Technip, 2006.

[68] Stéphane His, *Nouvelles Technologies de production de biocarburants*, Panorama IFP, 2007.

[69] Sunita Satyapal, John Petrovic, George Thomas, *Gassing up with hydrogen*, Scientific American, April 2007.

[70] Jeremy Rifkin, *The Hydrogen Economy*, Tarcher/Penguin Books, 2002.

[71] Stephen Boucher, *La révolution de l'hydrogène – Vers une énergie propre et performante*, Éditions du Félin, 2006.

[72] Christian Ngô et al. *Énergie*, Technologies du futur – Enjeux de société, Omniscience, 2005.

[73] EA Technology, *Review of electrical energy storage technologies and systems and of their potential for the UK*, Contract number DG/DTI/00055/00/00.

[74] Joseph J. Romm, Andrew A. Frank, *Hybrid vehicles gain traction*, Scientific American, April 2006.

[75] *La régulation intelligente pour sécuriser les réseaux électriques*, l'Usine nouvelle, 7 December 2006.

[76] Petroleum Reserves and Resources – Classification, Definition and Guidelines, SPE, AAPG, WPC, SPEE, Draft for industry review, September 2006.

[77] *Resources to Reserves*, Oil & Gas Technologies for the Energy Markets of the Future OECD/IEA, 2005.

[78] Jean-Luc Wingert, *La vie après le pétrole – De la pénurie aux énergies nouvelles*, Éditions Autrement, Paris, 2005.

[79] Xavier Boy de la Tour, *Le pétrole, au-delà du mythe*, Éditions Technip, 2004.

[80] Pierre-René Bauquis, *What future for extra heavy oil and bitumen: Orinoco case*, World Energy Council, 17th Congress.

[81] Albert Legault, *Pétrole, Gaz et les autres énergies – Le petit traité*, Éditions Technip, 2007.

[82] Jean Laherrère, Colin Campbell, *Oil and Gas Journal*, 14 July, 20, 2003.

[83] *Total, Rapport sociétal et environnemental* 2003.

[84] Marie-Françoise Chabrelie, *L'industrie gazière à l'horizon 2020*, Panorama, 2006.

[85] Marie-Françoise Chabrelie, *Le GNL, une commodité en devenir*, Panorama, 2006.

[86] Groupe de travail sur le charbon du Délégué interministériel du développement durable, *Charbon propre: mythe ou réalité*, 2006.

[87] Jean-Marie Amouroux, *Charbon chinois et développement durable*, 2007.

[88] Yves Bamberger, Bernard Rogeaux, *Quelles solutions des industriels peuvent-ils apporter aux problèmes énergétiques*, Revue de l'énergie, no. 575, January–February 2007.

[89] Henry Prévost, *Trop de pétrole – Énergie fossile et réchauffement climatique*, Éditions du Seuil, 2007.

[90] *CO_2 capture and storage in the subsurface – A technological pathway for combating climate change*, Collection Les enjeux des géosciences, IFP, ADEME, BRGM, 2007.

[91] Pierre le Thiez, Alexandre Rojey, *Le stockage géologique du CO_2: une vue générale*, Pétrole et Techniques, no. 449, March–April 2004.

[92] *Carbon dioxide capture and storage*, IPCC Special Report, IPCC Working Group III, Cambridge University Press, 2005.

[93] *Legal Aspects of Storing CO_2*, OCDE/IEA, 2007.

[94] Ron Pernick, Clint Wilder, *The Clean Tech Revolution – The Next Big Growth and Investment Opportunity*, Collins, 2007.

[95] *Climate change and a European low-carbon energy system*, EEA Report no. 1, 2005.

[96] Detief P. van Vuueren *et al.*, *Regional costs and benefits of alternative post-Kyoto climate regime*, RIVM Report 728001025/2003, National Institute of Public Health and the Environment, Bilthoven, 2003.

[97] Lester R. Brown, *Plan B 3.0 – Mobilizing to Save Civilisation*, Earth Policy Institute, 2008.

[98] Thomas L. Friedman, *Hot, Flat and Crowded - Why the World Needs a Green Revolution and How We Will Renew Our Global Future*, Allen Lane, 2008.

[99] Robert Bell, *The Green Bubble - Waste into Wealth: the New Energy Revolution*, Scali, 2007.

Web Sites

General sites

ADEME: www.ademe.fr
AIE (Agence Internationale de l'Énergie): www.iea.org
CEA: www.cea.fr
DOE (Department of Energy USA): IR www.doe.gov
IFP: www.ifp.fr
WEC (World Energy Council): www.worldenergy.org
MEDAD (Ministère de l'Écologie et du Développement durable): www.ecologie.gouv.fr
Ceren (Centre d'études et de recherches économiques sur l'énergie): www.ceren.fr
Institute for Energy and Environmental Research: www.ieer.org
Direction Générale de l'Énergie et des matières premières de l'Union Européenne: www.ec.
   europa.eu/energy
NEDO (New Energy and Industrial Technology Development Organization, Japan): www.
   neso.go.jp

Climate change

GIEC/IPCC: www.ipcc.ch
J.M. Jancovici's web site: www.manicore.com
MIES (Mission interministérielle de l'effet de serre): www.effet-de-serre.gouv.fr
United Nations Framework Convention on Climate Change: www.unfcc.int
Réseau Action Climat France: www.rac-f.org
Action carbone: www.actioncarbone.org
Combat Climate Change: www.combatclimatechange.org

Sustainable development – Environment

CITEPA: www.citepa.org
Earth Policy Institute: www.earth-policy.org
Greenpeace: www.greenpeace.fr
IFEN (Institut français de l'environnement): www.ifen.fr
World Resources Institute: www.wri.org
World Watch Institute: www.worldwatch.org
WWF (World Wildlife Fund): www.wwf.org
International Institute for Sustainable Development: www.iisd.ca
Centre population et développement: www.ceped.cirad.fr
Fondation Nicolas Hulot: www.nicolas-hulot.org

Energy efficiency

ADEME: www.ADEME.fr
Energy Star: www.energystar.gov
Association négawatt: www.negawatt.com
Batirbio: www.batirbio.org
Énergie-Cités: www.energie-cités.eu
Eurocities: www.eurocities.org
Fédération nationale des associations des usagers de transport (Fnaut): www.fnaut.asso.fr
Groupement des autorités responsables des transports (GART): www.gart.org
Centre d'études sur les réseaux, les transports et l'urbanisme (Cerfu): www.cerfu.fr
World Urbanization Prospects: www.esa.un.org
Groupe énergies renouvelables, environnement et solidarités (Geres): www.co2solidaire.
    org

Oil

AFTP: www.aftp.net
ASPO: www.peakoil.net
IFP: www.ifp.fr
TOTAL: www.total.com

Natural gas

Gaz de France: www.gazdefrance.fr
AFG: www.afg.asso.fr
UIIG (Union Internationale de L'Industrie du Gaz): www.gaz-naturel.ch
CEDIGAZ: www.cedigaz.org

Coal

ATIC (Association Technique de l'Importation Charbonnière): www.atics.fr
World Coal Institute: www.worldcoal.org
IEA clean coal centre: www.iea-coal.org.uk

Nuclear

CEA: www.cea.fr
AREVA: www.areva.com
World Nuclear Association: www.world-nuclear.org
SFEN (Société Française d'Énergie Nucléaire): www.sfen.org
ANDRA (Agence Nationale pour la gestion des Déchets Radioactifs): www.andra.fr

Renewable energies

ADEME: www.ADEME.fr
L'espace des énergies renouvelables: www.espace-enr.com
Observ'ER (Observatoire des énergies renouvelables): www.énergies-renouvelables.org
Comité de liaison énergies renouvelables (CLER): www.cler.org
Institut National de l'Énergie Solaire (INES): www.ines-solaire.fr
NREL (National Renewable Energy Laboratory, USA): www.nrel.gov

Biofuels

European Biofuels Technology Platform: www.biofuelstp.eu
Ademe: www.ademe.fr
IFP: www.ifp.fr

Hydrogen

AFH2: www.afh2.org
European Hydrogen and Fuel Cell Technology Platform: www.hfpeurope.org
International Partnership for Hydrogen Economy: www.iphe.net
International Association of Hydrogen Economy: www.iahe.org

CO_2 capture and storage

Club CO_2: www.clubco2.net
International Energy Agency GHG R&D Programme: www.ieagreen.org.uk
European Technology Platform for Zero Emission Fossil Fuel Power Plants (ZEP): www.
    zero-emissionplatform.eu
BRGM: www.brgm.fr
IFP: www.ifp.fr
CSLF (CO_2 Sequestration Leadership Forum): www.cslforum.org

Index

Page references in bold type refer to tables; those in italics refer to figures.

Energy and Climate: How to achieve a successful energy transition Alexandre Rojey
Copyright © 2009 Society of Chemical Industry